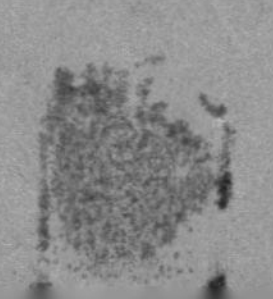

Cluster Analysis for Social Scientists

Techniques for Analyzing and Simplifying Complex Blocks of Data

Maurice Lorr

Cluster Analysis for Social Scientists

Jossey-Bass Publishers
San Francisco • Washington • London • 1983

CLUSTER ANALYSIS FOR SOCIAL SCIENTISTS
Techniques for Analyzing and Simplifying Complex Blocks of Data
by Maurice Lorr

433 California Street
San Francisco, California 94104
&
Jossey-Bass Limited
28 Banner Street
London EC1Y 8QE

Library of Congress Cataloging in Publication Data

Lorr, Maurice
Cluster analysis for social scientists

Bibliography: p. 202
Includes index.
1. Cluster analysis. 2. Social sciences—Statistical methods. I. Title.
HA31.3.L67 1983 519.5′3 82-49283
ISBN 0-87589-566-2

Manufactured in the United States of America

The paper in this book meets the guidelines for permanence and durability of the Committee on Production Guidelines for Book Longevity of the Council on Library Resources.

JACKET DESIGN BY WILLI BAUM

FIRST EDITION

Code 8314

The Jossey-Bass
Social and Behavioral Science Series

Special Adviser
Methodology of Social and Behavioral Research

Donald W. Fiske
University of Chicago

Preface

Cluster analysis is a generic term for a wide range of statistical techniques used to group objects, persons, stimuli, or concepts into homogeneous classes on the basis of their similarities. A cluster analysis results in clusters, types, classes, or groups. Because of their diverse origins the techniques are also categorized as typological analysis, numerical taxonomy, pattern recognition, and classification. Cluster analysis has many purposes: data reduction, the identification of natural grouping or types, the generation of useful classification schemes, and the testing of hypotheses.

Beginning about 1955 with the onset of the computer revolution, a veritable explosion has occurred in the development and publication of cluster-analytic techniques. More than 30 books and hundreds of articles have been published in a wide range of technical journals. During the initial period workers in different fields were quite unaware of comparable concepts and methods already developed by others. The founding of the Classification Society in 1970, and the publication of *Principles of Numerical Taxonomy* by Sokal and Sneath (1963), integrated the field and brought some consensus on common terms and concepts.

Application of cluster analysis to classification problems is evident in many fields:

- Behavioral and social sciences
- Biological sciences
- Engineering sciences
- Earth sciences
- Medicine
- Economics, political science, and business
- Library science and information retrieval

Despite this interest and flood of publication there is as yet no book on cluster analysis written at an elementary level for researchers and graduate students in the social and behavioral sciences. The present volume is designed for investigators and students who have completed, say, a couple of basic courses in correlation and one-way analysis of variance. While it does not provide a nonmathematical account of the techniques, a deliberate effort has been made to write at a fairly elementary level with repeated definition of technical terms and a glossary. Although use is occasionally made of matrix algebra to represent complex relationships difficult to express otherwise, the reader can skip these passages or examine the narrative formulations. It is not assumed that the reader knows how to use a computer or that he or she can read FORTRAN language. At the present day, at nearly every university, hospital, or business institution a computer programmer or trained graduate student is available to conduct the desired analyses. Thus the discussions presented here should enable a researcher to design a study, collect the necessary data, and select procedures appropriate to the problem.

Chapter One discusses the main goals of cluster analysis and then differentiates related procedures such as classification, identification, and discrimination. The chapter also presents a brief historical sketch of the development of clustering techniques. Chapter Two presents an overview of the process of cluster analysis and the various techniques. Chapter Three is concerned with the problems involved in measuring similarity and dissimilarity between cases on the basis of a set of attributes; measures of distance, correlation, and association are described and illustrated.

Chapter Four is devoted to a process called ordination by which entities are mapped within a low-dimensional attribute space. In this way the researcher is given a visual picture of clusters. Chapter Five outlines the single-level or nonhierarchical methods. Included are techniques for successive cluster formation and methods for partitioning data iteratively to form multiple clusters simultaneously. Chapter Six is devoted to the hierarchical techniques, which are classified as either agglomerative or divisive. The agglomerative methods proceed by successive mergers of the N entities into more and more inclusive clusters; the divisive techniques partition all the data into smaller and smaller subsets.

Chapter Seven reviews empirical studies of the validity of clustering algorithms and describes various criteria for ensuring accurate recovery of cluster structure. A method called Q analysis has often been used to find clusters or types; Chapter Eight examines the logic behind the procedures and offers some illustrations. Having established a taxonomic classification, the researcher can then use the subgroups for prediction; Chapter Nine deals with actuarial and other forms of prediction. Chapter Ten is concerned with hybrid models and overlapping clusters. Appendix A describes cluster-analytic computer programs and software; Appendix B reviews the rudiments of matrix concepts and operations. A glossary defines the technical terms.

I would like to acknowledge the crucial help of several individuals. Donald W. Fiske read the entire manuscript and made many helpful suggestions. Antanas Suziedelis generously reviewed most chapters. My daughter Nancy Lorr read chapters as they emerged and helpfully discussed obscure passages. I am also grateful to Catherine Wilmer for her patience with a burdensome script and her fine typing on the manuscript. Finally, I want to acknowledge the understanding and tolerance of my wife Joan Lorr, whose support made this work possible.

Washington, D.C. Maurice Lorr
June 1983

Contents

The Author

Maurice Lorr is professor of psychology emeritus and senior fellow of the Center for the Study of Youth Development at the Catholic University of America. He earned the B.A. degree and M.A. degree in psychology at the University of Illinois at Champaign-Urbana. He obtained the Ph.D. degree in psychology from the University of Chicago under L. L. Thurstone's chairmanship. For several years, Lorr worked as clinical psychologist at the Institute for Juvenile Research in Chicago. After serving as captain in the U.S. Army during World War II, he was appointed chief of research in the Division of Clinical Psychology at the Veterans' Administration Central Office. He then served as chief of the Outpatient Psychiatric Research Laboratory from 1953–1967, which was centered in the Washington, D.C. Veterans' Administration Hospital. In 1967 he was appointed professor of psychology and head of the personality psychology program at the Catholic University of America.

Lorr has been president of the Society for Multivariate Experimental Psychology and president of the District of Columbia Psychological Association. In 1980 he received the Award for Outstanding Scientific Contributions to Psychology from the Maryland Psychological Association. He is principal author (with C. J. Klett

and D. M. McNair) of *Syndromes of Psychosis* (1963). He is also editor of *Explorations in Typing Psychotics* (1966). Lorr is the principal author of the *Inpatient Multidimensional Psychiatric Scale* (IMPS) and the *Psychotic Inpatient Profile* (PIP), internationally known scales for rating psychotic patients. He is also coauthor (with R. P. Youniss) of the *Interpersonal Style Inventory* and (with D. M. McNair) of *Profile of Mood States*. He has published over 150 substantive and methodological papers.

Currently Lorr is advisory editor of *Multivariate Behavioral Research* and the *Journal of Clinical Psychology*. He is former advisory editor of the *Journal of Consulting and Clinical Psychology*. He is a fellow of Division 5 (Evaluation and Measurement) of the American Psychological Association and a member of Psychometric Society, the Classification Society, and the American Psychopathological Association.

Cluster Analysis for Social Scientists

Techniques for Analyzing and Simplifying Complex Blocks of Data

CHAPTER 1

Purpose and Uses of Cluster Analysis

Cluster analysis refers to a wide variety of techniques used to group entities into homogeneous subgroups on the basis of their similarities. Other names given the process are numerical taxonomy, pattern analysis, and typing. The end products of such analyses are called classes, types, groups, categories, or clusters. Since the number and nature of such groups are not known in advance, the clustering process is actually preclassificatory. In other words these techniques construct a classification scheme for unclassified data. Classification, identification or assignment, and discriminant analysis are key concepts closely linked to cluster analysis.

The term *classification* may refer to an end product or to a process. As an end product, a classification is a systematic scheme or arrangement of classes already known or established — such as Mendeleev's periodic table of the chemical elements. As a process, classification consists of sorting individual objects into well-known classes — voters, for example, can be classified as Republican, Democrat, Socialist, and Independent. *Identification* is the allocation of individual objects to established classes on the basis of specific criteria much as one identifies unknown trees by following keys in a leaf guide. In psychiatry and medicine the same process, called diagnosis, refers to the identification of a familiar disorder from the symptoms

presented by the patient. *Discriminant analysis* is a process undertaken to differentiate between groups formed on an *a priori* basis. The comparison might be between successful or unsuccessful salespersons, for example, or high, middle, and low social classes. The problem here is not to discover groups but to identify a set of characteristics that can significantly differentiate between the groups. The process allows one to allocate new cases to one of the groups with the least amount of error. In contrast, the clustering process generates a new categorical scheme or recovers groups within a mixture of several populations.

One more differentiation in terms is needed. In the technical literature the name *cluster analysis* sometimes refers to two quite different procedures. Prior to 1955 cluster analysis represented a method of grouping variables instead of objects into subsets in order to define dimensions or factors. Tryon (1939) developed several such techniques to avoid using *factor analysis,* a relatively complex statistical procedure that was laborious and time-consuming before the computer revolution. Tryon conceived of his cluster analysis as a poor man's factor analysis. In this book, however, clustering refers to the formation of homogeneous subgroups of entities such as persons, plants, ships, or pots. An example of such an analysis is Miller's (1969) study of 48 common nouns grouped into five clusters referring to living things, nonliving things, quantitative terms, social interactions, and emotions. Another illustration is Paykel's (1971) analysis of 165 depressed patients. Using symptom ratings and historical variables, he grouped his cases into four clusters: the retarded psychotic, the anxious, the hostile, and the young depressive.

In factor analysis, interest centers on the similarity of the variables (attributes). The aim is to identify a small number of dimensions that can account for individual differences on the various measures or attributes. Thurstone, for example, analyzed relations among 57 cognitive tests and identified seven factors interpreted as representing abilities to solve verbal, numerical, spatial, and other tasks. Another example is found in Osgood, Suci, and Tannenbaum's (1957) study in which 100 subjects rated 20 concepts on 50 scales of semantic judgment. The factor analysis of the scales

disclosed three underlying dimensions of affective meaning: evaluation (good-bad), potency (strong-weak), and activity (active-passive).

The Data Matrix

The input in a cluster analysis consists of a *raw* data matrix — that is, a rectangular array of numerical entries. The cell values may consist of nominal, ordinal, or interval measurements of the objects or the variables. The set of objects is represented in the N rows; the set of variables is represented in the n columns. The (i,j)th cell entry represents the value of an object i on variable j. The objects may be people, words, stimuli, plants, or concepts. The variables are attributes or characteristics of the set of objects. The complete row of values is called the object's *profile.* The next step in most cluster analyses is to transform the data matrix into a square $N \times N$ matrix of interobject similarity or dissimilarity measures. Given that the number of groups is not known, the problems of cluster analysis are threefold: to choose a measure of interobject similarity, to select a method for forming subgroups once the indices of similarity/dissimilarity have been obtained, and then to decide on the number of subgroups present in the data or to construct a hierarchical arrangement. This brief summary of the sequence of steps in a cluster analysis will be further elaborated on in the next chapter.

Aims of Cluster Analysis

The general objective of cluster analysis is to *partition* or subdivide a set of objects into homogeneous subgroups or into a hierarchical arrangement of homogeneous subgroups. Stated more specifically, these techniques are designed to:

- Identify natural clusters within a mixture of entities believed to represent several distinguishable populations
- Construct a useful conceptual scheme for classifying entities
- Generate hypotheses within a body of data by discovering unsuspected clusters

- Test hypothesized classes believed present within a certain group of cases
- Identify homogeneous subgroups characterized by attribute patterns useful for prediction

The first aim refers to a resolution of what is known to statisticians as the *mixture problem.* This view assumes that the data set to be analyzed consists of samples from several populations but the number of populations and their distribution parameters (means and dispersions) are unknown. The task is to resolve the mixture and identify the populations and their characteristics by using sample information. The clusters that emerge are sometimes labeled *natural* to emphasize that they can be used for many purposes. More will be said later to contrast general with special-purpose clusters.

The second purpose of clustering relates to a problem that often confronts social or behavioral scientists. Once a large mass of data has been collected on numerous cases using many measures, the problem is one of data reduction. By applying clustering techniques, information regarding a sample of N cases can be reduced to information concerning a smaller number of g groups. Construction of a taxonomy simplifies the observations with a minimal loss of information. It also provides a taxonomy that can contribute significantly to an understanding of the problem studied.

The third purpose, the production of hypotheses regarding subgroups that may be present in the data, motivates many analysts. The researcher would be wise to examine the data for heterogeneity, however, before clustering techniques are applied. If the multivariate distributions are unimodal and symmetric, the sample may be homogeneous. Further, the groupings that do emerge must be replicated on a new set of observations since the information from which the hypotheses were derived cannot be used as its own evidence. Goldstein and Linden's (1969) study of alcoholics by means of the MMPI offers an illustration. When two large independent samples of hospitalized alcoholics were analyzed on the basis of their MMPI scale profiles , four alcoholic types were identified. The same four types were replicated in the second sample.

The fourth aim of testing for conjectured groups can be illustrated by a study conducted by Boyce (1969). The bases of the

comparison were measures on the skulls of four groups of hominoids: chimpanzees, orangutans, gorillas, and homo sapiens. Several cluster analyses of the 20 skulls indicated that these groups could in fact be differentiated.

The fifth purpose of clustering is to formulate a taxonomic scheme of potential use in prediction. The approach assumes that the individuals within homogeneous subgroups will also be similar with respect to predictor dimensions. Arguing that the best predictor of future behavior is past behavior, Owens (1968) proposed that subgroups of individuals should be formed on biodata or life-history variables so that individuals within a group would have similar life histories. When Feild, Lissitz, and Schoenfeldt (1975) administered a 375-item biodata blank to large samples of men and women, 13 scores of the subjects were grouped by a hierarchical procedure to determine 23 subgroups. Subgroup information was found to contribute significantly in predicting a set of criteria.

Uses of Types

In order to state general propositions regarding individual objects we must sort them into classes. In fact classification seems to be a process necessary for *all* thinking and language since it involves the generation of concepts that reflect uniformities in the subject under study. Scientific constructs are simply general concepts upon which a particular science develops an understanding of the phenomena within a field. A *taxonomic hierarchy* organizes the various classes into comprehensible arrangements.

When a cluster analyst asks what a cluster means, what it represents, or what it implies, he or she is asking what construct or *type* underlies the data. The *cluster* represents an operational definition of a grouping of entities found in empirical data. A type is a construct inferred from the cluster. Thus although the terms are used as synonyms, here I will use the term cluster to refer to a grouping defined by an algorithm. The word type will be used in a more limited way to refer to an explanatory concept. Comparable constructs in the social and behavioral sciences are such concepts as attitude, trait, motive, or value.

At the practical level types are useful insofar as they facilitate

communication. Types are easy to identify, to remember, to report, and to differentiate from other objects in a domain. Categorization also provides direction for reactions to the environment. We then know in advance how to act and what behavior is appropriate or inappropriate. In this sense a psychopath is treated differently from a person labeled a paranoid or a depressive. Similarly, people respond differentially to conservatives and liberals, radicals and reactionaries. Communication about objects (friends, patients, diseases, stimuli) is enhanced because the label brings to mind a large set of characteristics that call for attention or action.

Typing also helps reduce the complexity of a set of data. Comprehension is facilitated when individual cases can be sorted into smaller sets of cohesive categories or into a hierarchical arrangement. Categorization leads to parsimony and greater efficiency. It can also be argued that typologies increase our ability to predict an attribute of interest (called a criterion or dependent variable). The relation of a test to a criterion may be very different within a type as opposed to between types. As Forgy (1965) argues, "A typology can reflect a fact of nature, that there are actually discrete, separate subtypes of individuals within a larger sample." The natural cluster represents such a summarization.

Natural vs. Special-Purpose Types

The attributes that form the basis of a classification must represent a selection from all possible characteristics. The selection depends on our purpose: To study voting behavior, people are questioned on their political beliefs; to select students for university admission, people are assessed on their cognitive skills. In other words a group of persons can be composed by occupation, by nationality, by race, by personality, as well as in innumerable other ways. Clearly no single all-embracing classification is possible. One reason is that the basis of a classification depends upon the researcher's interest and purpose. Another reason, equally fundamental, is that similarity is not a general characteristic. It is always necessary to specify the attributes on which a set of entities are compared.

Types that emerge from clustering can be differentiated into

general-purpose classes and special-purpose classes. General-purpose clusters are designated *natural classes* by biologists (Gilmour, 1937). A number of criteria have been proposed as a basis for calling some classifications more natural than others. Clusters are natural if a large number of propositions can be stated regarding their members. Moreover, natural classes convey a high content of information and can be used for many purposes. Hempel (1952) refers to such characteristics as being of systematic import. Classification of individuals by eye color, year of graduation from high school, or first letter of the surname produces special-purpose groups. Similarly, a classification of current members of the Maryland State Senate by voting record would be useful but temporary and lacking generality. On the other hand, a classification of depressive disorders into subtypes could be broadly useful for predicting type of treatment and duration of illness and would suggest an etiological basis. Tryon's (1955) identification of social areas in the San Francisco Bay area as "Skid Row," "Exclusive Area," "Working Class Section," and "Little Italy" is useful because it throws light on American cities generally.

A Brief History of Classification

The systematic grouping of objects on the basis of common properties dates back to Aristotle and the Greeks. To Aristotle, taxonomy consisted of efforts to discover whatever properties define the essence of a class or *taxon.* These properties are then necessary, both singly and jointly, to justify membership in the class. All characteristics must be present for one to say that the object is a member of the class—for example, the presence of all symptoms, fever, cough, and a rash, suggests a diagnosis of measles. Once a classification is established, an object is first assigned to the larger group (the genus) to which it belongs by virtue of possession of essential characteristics of that group. The object is then assigned to the subgroup (the species) on the basis of the essential characteristics of the species and so on.

Historically the early taxonomic schemes consisted of efforts to discover whatever characteristic defined the essence of a class of objects. This approach was followed, at least in part, by Linnaeus the

Swedish botanist and taxonomist in his classifications of plants (*Genera Plantarum,* 1737), animals, and minerals. His schemes for the classification of botanical specimens had widespread impact on other fields.

Sneath and Sokal (1973) credit Adanson (1763), a botanist, with the idea that natural taxa should be based on the concept of similarity, which is measured by taking many characters into account. Such taxa or types are *polythetic*—that is to say, based on multiple characteristics of the objects compared. The earlier taxonomic systems favored *monothetic* types formed by a divisive technique requiring the presence of a single specified binary attribute for membership. Some modern biological taxonomists (for instance, Sneath and Sokal) view taxa as subgroups of organisms that manifest a fair number of shared properties. Attributes of a class need only be correlated with category membership. Arrangements made on the basis of overall similarity are said to reflect *phenetic* relationships. The process of grouping taxa (individuals or groups) by numerical method is called numerical taxonomy by these biologists.

In many respects cluster analysis was given its major impetus in the 1950s by the development of high-speed computers and the rapid appearance of clustering algorithms. The biologists active in proposing clustering techniques were Sneath (1957) and Sokal and Michner (1958); in psychology McQuitty suggested very similar procedures (1957, 1961). Another event of importance was the publication of *Principles of Numerical Taxonomy* by Sokal and Sneath (1963). Although directed toward biologists the book reviewed much of the literature, presented a clear outline of the various steps taken in a cluster analysis, and described many clustering techniques available at that time.

The 1970s saw publication of five new texts and a revision of *Numerical Taxonomy* by Sneath and Sokal (1973). Tryon and Bailey (1970) published *Cluster Analysis,* a text describing a package of computer programs called the BCTRY system for clustering variables and objects. *Mathematical Taxonomy,* by Jardine and Sibson (1971), is a highly technical and mathematical work directed primarily at biologists. Anderberg's *Cluster Analysis for Applications* (1973) is concerned with clustering variables as well as entities; it is thorough in its coverage, well illustrated, and critical in its approach.

Clustering Algorithms by Hartigan (1975) is the first text presented from the statistician's point of view: A wide range of procedures are described and illustrated, and FORTRAN programs are presented for the various algorithms. The notation, however, intended for the computer programmer, may be difficult for the ordinary researcher to follow. Everitt's *Cluster Analysis* (1974) is thorough in its coverage, but readers are expected to be fairly sophisticated in their knowledge of the statistics of multiple variables.

Clifford and Stephenson's *An Introduction to Numerical Classification* (1975) is intended for biologists. Duran and Odell's *Cluster Analysis: A Survey* (1974) seems directed at the statistically sophisticated economist. Recently Spath published *Cluster Analysis Algorithms* (1980) in Germany. *Classification and Clustering* (Van Ryzin, 1977) presents the proceedings of an advanced seminar conducted at the University of Wisconsin by the Mathematics Research Center; included are 16 papers by well-known mathematicians and statisticians such as Hartigan, Kruskal, Rao, and Solomon. The recently published symposium volume *Classifying Social Data: New Applications of Analytic Methods for Social Science Research* (Hudson, 1982) grew out of a workshop on classification procedures held at Charleston, South Carolina. The first six chapters represent cluster and factor-analytic studies of social data; the remaining seven chapters present innovative methodological approaches such as blockmodeling and empirical comparisons of cluster techniques.

There are a number of useful critical reviews of cluster analysis. Still relevant are early reports by Ball and Hall (1965) and another by Fleiss and Zubin (1969). Cormack's (1971) review of classification is probably the most sophisticated, thorough, and critical of all that have appeared; virtually all aspects of theory and method are examined. Bailey (1974) presents a useful survey of cluster analysis in *Sociological Methodology 1975*. Blashfield and Aldenderfer (1978) report a useful review of the cluster analysis literature of the 1970s.

Summary

This chapter has described the purpose and development of clustering techniques. The process of cluster analysis must be differ-

entiated from related procedures such as classification, identification, discriminant analysis, and factor analysis. The techniques of cluster analysis have many uses: finding natural clusters, constructing conceptual schemes for classifying entities, generating hypotheses for future confirmation, testing conjectured classes, and finding taxonomies potentially useful in prediction. To the scientist cluster analysis offers a means to construct conceptual schemes for organizing empirical data in a field, to assist in communication, and to reduce the complexity of a set of data. Natural or general-purpose classes are differentiated from special or limited purpose categories. Finally, the chapter has sketched the historical background of classification and listed some of the currently available sources of algorithms.

The next chapter presents an overview of the process of cluster analysis. In this way the potential user will become aware of the sequence of steps involved in the discovery or recovery of groupings of interest. Subsequent chapters will deal with the details of these steps.

CHAPTER 2

Process of Analysis

This chapter is concerned with the process of cluster analysis and its various steps. Given a sample of N entities described or measured on each of n variables, the collected data are assembled in a data matrix of N rows and n columns. The problem is how to compare the entities and how to group them into subsets on the basis of similarity. This chapter describes the sequence of critical steps involved in discovering or recovering groups. Each step is explicated at some length to illuminate the *process* of cluster analysis. Of special interest are the choice of an index of similarity, choice of a structural model, and choice of a cluster search technique.

The Process

Clustering is the grouping of entities into subsets on the basis of their similarity across a set of attributes. The initial data are commonly assembled in a two-way table. The rows represent entities such as people, concepts, stimuli, or objects. The columns represent attributes such as attitudes, skills, interests, and personality traits. A sociologist might want to study prisoners in terms of the offenses they have committed (burglary, larceny, rape, or assault, for example). A psychiatrist might be interested in grouping hospitalized patients in terms of their symptoms and behavior. An occupational specialist might wish to group workers from various industries into

job families. These are a few examples of problems that call for cluster-analytic techniques.

Once the investigator has decided that a cluster analysis might reveal suspected subsets, a study plan is needed. What kind of sample should be collected? What kinds of attributes should be measured or recorded? What measure of similarity should be chosen to compare the entities? Once the similarity measures have been computed, there are other questions: What cluster search techniques should be chosen? Should the set of entities be partitioned into separate clusters or should a hierarchical arrangement be sought? These are some of the problems encountered in conducting a cluster analysis.

A well-designed study usually involves a sequence of steps such as the following:

1. Select a representative and adequately large sample of entities for study.
2. Select a representative set of attributes from a carefully specified domain of similarity.
3. Describe or measure each entity in terms of the attribute variables.
4. Conduct a dimensional analysis of the variables if these are very numerous.
5. Choose a suitable metric and convert the variables into comparable units.
6. Select an appropriate index and assess the similarity between pairs of entity profiles.
7. Select and apply an appropriate clustering algorithm to the similarity matrix after choosing a cluster model.
8. Compute the characteristic mean profiles of each cluster and interpret the findings.
9. Apply a second cluster analytic procedure to the data to check the accuracy of the results of the first method.
10. Conduct a replication study on a second sample if possible.

Choice of Attributes

A careful definition of the attribute domain is of considerable importance. Similarity, as we have noted, is not a general quality. When entities are compared on independent attribute dimensions,

entities similar on one dimension need not be similar on other dimensions. That is, there is no single way to categorize people or other entities. People who are alike with respect to one set of attributes are not necessarily more alike on other attributes than people in general. People may be alike in political attitude but very different in food preference, body type, and personality style. Thus the notion of similarity has meaning only with respect to a specified set of attributes.

Having specified the domain of similarity, the investigator must set up a procedure for selecting a broadly representative sample of attributes. All pertinent entity differences should be represented. When relevant variables are left out of an analysis, some groups will remain undifferentiated from groups that are adequately defined. If vocational interest types are sought, for instance, it is important to include all the interests likely to separate the hypothesized types. If criminal types are being studied, a wide range of both blue-collar and white-collar crimes ought to be tested. If a domain is incompletely represented, otherwise distinctive subgroups are likely to be confounded.

Choice of Entities to Be Studied

The objects of a cluster analysis typically represent a sample from a much larger population. As such they are referred to as subjects, cases, taxonomic units, observations, or data units. The sampling process should ensure that all data units have an equal opportunity to be selected as part of the sample. Any types believed to be present in the data should be represented in the sample in proportion to their relative size. Also relevant here are extraneous but influential sources of variance such as age, sex, social class, education, and income. The critical consideration throughout is to make sure that all types of individual variation are represented.

Dimensional Analysis

Factor analysis is a statistical procedure in wide use in the behavioral and social sciences. It is designed to isolate and identify the main sources of individual variation in the data. These sources,

called *factors,* are best conceived as dimensions of individual difference. More specifically the method is used to reduce the number of variables to a parsimonious set, to generate hypotheses regarding the number and kinds of dimensions present, and to test or confirm some hypothesized factor structure.

It is generally useful to conduct a dimensional analysis of the descriptive variables if they are too numerous, highly redundant, or complex. Suppose data were collected on a sizable sample by means of a lengthy inventory of 400 questions. A factor analysis could easily reduce the 400 variables to 15 factors. The *factor scores*—that is, the measures based on the 15 factors—would be more easily understood, more easily reported, and more reliable than individual variables.

Factor scores are more reliable than single variables because they are weighted linear composites of variables that best define the factor. In fact, the greater the number of positively correlated measures that are combined into a summary score, the more reliable the composite (Cronbach and Gleser, 1953). Reliability increases with length because the true component of a score is proportional to the number of equivalent elements that contribute to it. It follows, then, that factor scores representing composites tend to be more reliable than single variables.

Another benefit that results from a dimensional analysis of a large number of variables arises in the process of interpreting clusters. The usual practice is to describe each cluster in terms of the mean scores of members of the cluster. Since it is obviously cumbersome to report scores on several hundred variables, the usual practice is to identify the group in terms of a few characteristics that best describe it. A profile of 15 scores is much easier to report—and is certainly more readily comprehended—than a profile of 400.

Choice of Metric

In most cluster analyses the descriptive measures vary in dispersion and in the units expressed (the *metric*). In studying body types, for example, the variables may be height expressed in inches, weight expressed in pounds, and muscularity rated on a five-point

scale. Somehow these measures must be made comparable in the units expressed in order not to give implausible weights to some of the variables. Consider the variables in a demographic study. These may be age expressed in years, sex as a binary variable, level of education expressed in grades, and religion categorized into five nominal scales. The usual procedure is to standardize all dimensional attributes so that each scale has a mean of zero and a standard deviation of 1. Properties not capable of further subdivision (such as male or triangular) can be treated as present or absent—that is, as all-or-nothing properties—and then standardized (Guilford, 1965).

There are four *scale types* likely to be encountered in conducting a cluster analysis: nominal, ordinal, interval, and ratio scales. *Nominal* (or classificatory) *scales* are the most rudimentary. When numbers or symbols are used to identify groups to which objects or persons belong, they constitute a nominal scale—for example, the psychiatric system of diagnostic categories. *Scaling* is a process of partitioning the collection of objects into mutually exclusive subsets. The relation between objects is one of equivalence. Thus the only transformation permissible is one-to-one—that is, symbols or numbers may be interchanged providing the transformation is systematic and complete.

Ordinal scales represent consistent rank orders. Objects within successive categories differ by being greater than, or less than, the preceding. Any order preserving transformation will leave an ordinal scale unaltered. Thus it does not matter which numbers are given members of a rank order so long as the higher number is given to members of the class that is "greater than."

Interval scales are characterized by equal units but no zero points. Many psychological test scales satisfy the conditions of an interval scale. Interval scales have all the properties of an ordinal scale, but in addition they specify a distance between any two objects. A permissible transformation must preserve not only the order but also the relative difference between the objects. Distances and correlations can be computed with interval scale data. *Ratio scales,* which possess equal intervals and true zero points, are exemplified by length, weight, and time. Ratio scales are seldom encountered in the social sciences.

Choice of a Similarity Index

Similarity represents an organizing principle by which objects are classified, concepts are formed, and generalizations are made. In the last chapter we saw that we cannot create a classification without selecting attributes because similarity is not a general characteristic. The conventional theoretical model for similarity is geometric. Objects are represented as points in some attribute space such that the observed dissimilarities between objects correspond to the metric distance between the respective points. Tversky (1977), however, has questioned the metric and dimensional assumptions that underlie this geometric representation. He has presented empirical evidence for asymmetric similarities whereas the geometric representation assumes that the similarity relationship between points is symmetric. This means that the distance between *A* and *B* equals the distance between *B* and *A*. He argues that similarity should not be treated as a symmetric relation. We say "the portrait resembles the person" rather than "the person resembles the portrait." We say "the ellipse is like a circle," not "a circle is like an ellipse." Thus we tend to select the more salient stimulus as a referent and the less salient stimulus as the subject. The approach taken here is that since departures from the symmetric relationship in similarity data are generally small, symmetric measures do provide a useful first approximation to the discrete structures that are represented in the relation between objects.

Choosing a measure of similarity or difference between entities and attributes involves several problems. The first relates to the type of attribute employed. Nominal scales call for matching coefficients. Truly ordinal scales suggest the use of ordinal or rank order coefficients. Interval scales indicate the use of distance functions or correlations.

When the attributes are measured on interval scales, it becomes important to consider three aspects of the set of numerical values (scores) that characterize the entities. Each row of values in the $N \times n$ data matrix $\mathbf{X}$ defines the *profile* of an individual entity across attributes. The distance between any two profiles may then be computed as the square root of the sum of squared differences in scores between the two entities over the n variables. Here X_{ij} denotes

the score of entity i on test j in

$$D^2 = \Sigma(X_{ij} - X_{hj})^2$$

It can be shown (Cronbach and Gleser, 1953) that D^2 is composed of three components: elevation, scatter, and shape. The *elevation* of a profile is the mean level of the scores. The *scatter* represents the range or variation in the scores. *Shape* refers to the configuration of scores—that is, which are high and which are low. If the ranking of the scores in the two profiles is similar, the profile shapes too are very similar.

If the similarity between profiles is assessed by a distance function such as D, all the components of information present in the profiles are represented. But if the elevation is subtracted from each of the raw scores by conversion to deviation scores, then this source of information is of course lost. Likewise if the scores are standardized, information concerning scatter will be lost. The cluster analyst should therefore consider carefully the information needed when it comes to choosing a distance measure. The details of the problem are examined in the next chapter.

Another frequently used measure of similarity between profiles is the *correlation coefficient.* When profiles are correlated, the entity means are subtracted and each deviation score is divided by the scatter. It follows that a correlation coefficient does not reflect information regarding the profile elevation or its scatter. So long as the shapes of the two profiles are similar, the correlations will be high. Thus the investigator must weigh the advantages and disadvantages of each index.

Choice of Structural Model

There are two aspects to choice of a structural model. First the analyst should decide what kind of cluster he or she expects to find in the data. It may be either compact (spherical or ellipsoidal) or extended (serpentine). A second choice is between a hierarchical treelike arrangement of nested clusters or a nonhierarchical discrete set. The former usually reflects a developmental sequence, while the latter does not.

An intuitive understanding of a cluster can be gained by considering each entity as though it were a point in n-dimensional space. Visualize again the data matrix of N rows and n columns. Each of the n variables in the columns may be represented by one axis in the n-dimensional space. Each row gives the variable values for one of the entities. Thus the values can be regarded as the coordinates of the entity in attribute space. Some regions of this space will be densely populated while others will contain relatively few points. A cluster may then be visualized as a region of high density separated from other dense regions by a low-density area. There appear to be two kinds of clusters: compact and chained. Members of a *compact cluster* are characterized by high mutual similarity; all are located within a circumscribed distance of one another. In general the swarm of points is roughly circular or spherical in shape. A compact cluster may be defined as a subset of entities in which each member is more like *every* other member than it is like entities in any other cluster. The relationship that characterizes members is symmetric. Entity pairs are consonant-dissonant, like-different, near-far, or matched-unmatched. This proximity relationship is measured in terms of the absolute difference between entities (distance-disparity).

A *chained* (or *connected*) *cluster* is a subset of entities in which every member is more like *one* other member of the type than it is like any other entity not in the cluster. The swarm of points tends to be long and straggly, serpentine, or amoeboid. The extended nature of chained clusters suggests that entities are linked by asymmetric

Figure 2-1. Compact and Chained Clusters.

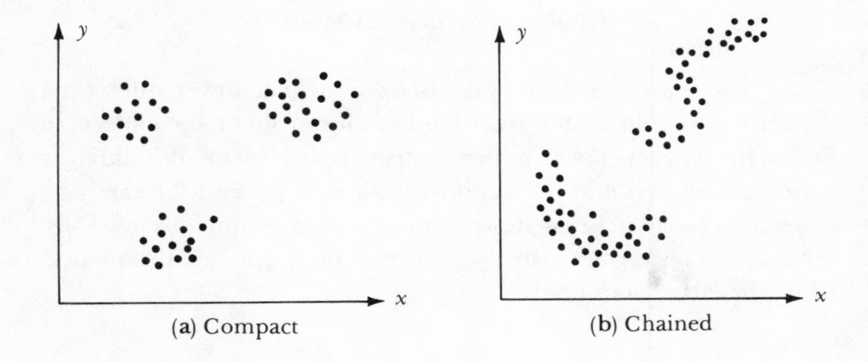

(a) Compact (b) Chained

Figure 2-2. Hierarchical Tree Structure.

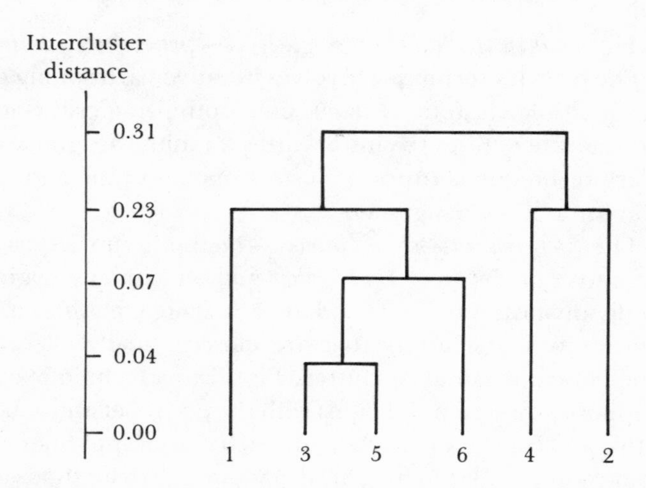

order relationships of greater than, dominant over, preferred to, or chosen over. Such ordinal relations are found in sociometric choices, communication networks, preferential orders, and competitive game rankings.

The *hierarchical scheme* is commonly represented as a *tree,* beginning at the branches and combining until the trunk is reached. Suppose you are given *N* entities and told to create a hierarchy. Treating each entity as a cluster, you merge the two that are closest. At the next level you merge the two closest and so on until all are combined at the trunk. Any two clusters are either disjoint or one includes the other. Thus a hierarchy is a nested set of clusters in which each level is assigned a rank. As we shall see, the defining relations between entities may be different—as in the compact and chained types. When the entity relationships are asymmetric and transitive (that is, ordinal), the clusters represented are continuously connected chains. Figures 2-1 and 2-2 illustrate the two kinds of clusters and the hierarchical tree structure.

Choice of a Cluster Search Method

The main clustering techniques are applied in conjunction with a matrix of similarity indices for each pair of entities. Usually these are distances or correlations. Clustering techniques are

broadly classifiable into nonhierarchical methods and hierarchical methods.

The *nonhierarchical*—or *single-level*—procedures are of two kinds. The primary technique involves iterative partitioning of entities into multiple clusters. Usually some optimizing criterion is applied to relocate entities to clusters after an initial assignment. The secondary technique is to form clusters one at a time and without iteration for a better assignment.

The *hierarchical*—or *multilevel*—methods can be classed as agglomerative or divisive. The *agglomerative* technique begins with all N individual cases or units and at each stage combines together the two entities or clusters that are closest; finally all cases are combined into one family or cluster. The *divisive* technique operates in the opposite direction: It begins with the entire set and subdivides it into two and continues to subdivide each cluster into finer subsets. The four most popular hierarchical agglomerative methods according to Blashfield (1976b) are known as *single-linkage, complete-linkage, average-linkage,* and *minimum-variance* technique. These methods differ primarily in terms of the rules by which entities or clusters are combined or linked. The details of these processes are presented in the ensuing chapters.

Biologists tend to rely on the hierarchical techniques because of their interest in evolutionary development. Although the nonhierarchical partitioning methods that yield single-level sets of clusters have greater appeal to social and behavioral scientists, the hierarchical methods are often preferred for several reasons. One reason is that generally poor performance has been obtained when nonhierarchical methods such as k means are applied without providing good starting points. A second reason is that criteria have been developed for determining the level in a hierarchy at which there is an optimum number of clusters present. A third basis for choosing a hierarchical method is that a developmental arrangement is expected and sought.

Summary

This chapter was concerned with the process of cluster analysis and its various steps. A representative sample of entities is selected and described in terms of a set of attributes of interest to the

researcher. The measures are commonly reduced in number by a dimensional analysis. The attribute variables are then combined and converted into a common metric. The similarity of entity profiles on the reduced set of attributes is assessed by some suitable index. A cluster analytic procedure, either hierarchical or partitioning, is applied to the matrix of similarity measures. After establishing the number of clusters present, the mean profile of each subgroup is determined. The accuracy of the results is checked by applying a second clustering procedure. Finally a replication study is conducted to establish the stability of the results.

Chapter Three deals with the problems that arise in measuring the similarity between profiles. Four levels of measurement are delineated. A broad distance function is presented and shown to relate to common indices of correlation, and various measures of association applied to categorical variables are described.

CHAPTER 3

Measuring Similarity Between Profiles

To cluster entities into homogeneous subsets, we must compute some measure of similarity or difference between each and every other entity. This chapter discusses the problems involved in measuring similarity between profiles. First we consider four scale types that affect the treatment of the profile elements. Then a general concept of distance is presented along with its components of elevation, scatter, and shape. After analyzing several distance measures and relating them to correlations, we shall evaluate the use of correlation coefficients as alternative measures and discuss several coefficients in detail. Finally, we shall consider some measures of association commonly used with categorized variables.

Variables and Scales of Measurement

A common assumption is that all variables are of a single type. Typically, variables are assumed to represent continuous scales with equal intervals. In the real world, however, the attributes and properties by which entities are described are a mixture of scales: nominal, ordinal, interval, and ratio. There are also at least three kinds of variables: continuous, discrete, and binary.

A *continuous variable* has an uncountable range. Age, height,

and test scores are continuous variables because they can have an infinite number of closely spaced in-between values. In each of these variables the number of possible values is much larger than the number actually in a particular scale. A *discrete variable* has a finite or countable range. Such a variable takes only point values such as 0, 1, 2, 3, and so on—the number of children, for example, or the number of responses to a question.

A *binary* (or dichotomous) *variable* is a special kind of discrete variable; it has only two values. Examples are sex, yes-no or true-false answers, and presence versus absence of a quality. It is also useful to distinguish between dichotomies and unordered polytomies, which consist of three or more mutually exclusive and exhaustive categories such as religion or occupation. Most techniques for analyzing association data lend themselves to binary-coded (0 – 1) variables but not to polytomies. But polytomies can be recoded into a set of arbitrarily assigned dichotomous *dummy variables* as 0 or 1. A dummy variable is a binary coded vector in which members of a group are coded 1, while nonmembers are coded 0. The necessary and sufficient number of vectors to code group membership is one less than the number of groups. We must create, for k-groups, $k - 1$ vectors (x_1, x_2, x_3) (Cohen and Cohen, 1975). Let us say, for example, that a person's religion can be classified into four categories: Catholic, Protestant, Jewish, Islamic. A Catholic would be coded 1 on dummy variable 1, and would be coded 0 on dummy variable 2 (Protestant) and 3 (Jewish). A Protestant would be coded 1 on dummy 2 and 0 on the other three variables. A Jewish subject would be coded 1 on dummy 3 and 0 on the other two. The last category of Islamic would receive a value of 0 on all $k - 1$ dummies. A fourth vector would be redundant since it would provide no additional information beyond what is contained in vectors x_1, x_2, and x_3; Islamic is represented implicitly by 0, 0, 0. In general a k-category polytomy can be represented by $k - 1$ dummy variables with the last categorized 0 on all $k - 1$ dummies.

A number of other terms need definition. An *entity* refers to a thing possessing certain properties. An *attribute* is a property capable of further division—in other words, it is a quantitative variable. Properties of objects not capable of further division are called *qualities.* The terms attribute, dimension, and continuum can be

used more or less interchangeably. An attribute is unidimensional if its degrees can be represented by a single rank order. In fact quantitative continua are referred to as *ordered* variables whereas qualitative variables, such as eye color and nationality that differ only in kind, are described as *unordered.* Let us now define the scale types and indicate their properties.

A *nominal scale* only distinguishes between objects or classifies them. When numbers or symbols are used to identify classes to which the entities belong, the symbols define the nominal scale. Examples are numbers assigned to football players, race, and psychiatric diagnosis. In a nominal scale, the scaling operation partitions a set of objects into mutually exclusive subsets. The relation between objects is one of equivalence—that is, members of a set must be equivalent on the property being scaled. If the variable is X and two objects are A and B, then either $X_A = X_B$ or $X_A \neq X_B$. The subsets may be represented equally well by any set of symbols, which may be interchanged providing the exchange is consistent and complete. The only descriptive statistics admissible are those that would be unchanged by one-to-one transformations. These are the mode, frequency counts, and the contingency coefficient C.

An *ordinal scale* reflects a rank order among the objects. In addition to distinguishing objects on the basis of equivalence, as in nominal scales, ordinal scales incorporate the relation "greater than" (>). Examples of ordinal scales are military rank, social status, and IQ level. It does not matter what numbers are assigned to ordered objects as long as the higher number is given to members of the class that is "greater than." Admissible statistics include the median and the rank order correlation, such as Spearman's or Kendall's tau (Siegel, 1956).

An *interval scale* is characterized by equal units of measurement. It assigns a meaningful measure of the difference between two objects. In addition to all the characteristics of the ordinal scale, the interval scale provides a distance between any two objects enabling us to say that A is so many units different from B on variable X. Temperature is an example of an interval scale. A linear transformation is permissible because it not only preserves the ordering of objects but also the relative difference between them. The information in the scale is not affected if each number is multiplied by a

constant or if a constant is added to each unit. All the familiar parametric statistics—such as means, standard deviations, and correlations—may be computed with interval scale data. The conversion of raw scores to standard scores with a mean of zero and a standard deviation of 1 is an example of a linear transformation.

A *ratio scale* is an interval scale with a true zero point as its origin. Since the ratio of any two scale points is independent of the unit of measurement, it is possible to say that a is X_a / X_b times greater than b. Length, time intervals, and loudness (sones) are examples of ratio scales, which are extremely rare in the social sciences. All ratio scales can be treated as interval scales.

Measurement Scales and Statistics

Some elementary statistics books state that different statistical procedures require the use of specific measurement scales (nominal, ordinal, interval, ratio). This view was first suggested by Stevens (1946) and later used by Siegel (1956), Blalock (1979), and others. Stevens specified the statistical measures for use with each scale: Nonparametric procedures are appropriate with nominal and ordinal scales, whereas parametric procedures are required for interval and ratio scales. He contended that measurement scales are models of entity relationships. The more the measurement model deviates from the properties of the objects measured, the less accurate the statistics. Since many social science measures are either ordinal or approximations to interval scales, it may be inappropriate to compute correlations and distances between entities.

An opposing view is held by many researchers such as Lord (1953), Burke (1953), Anderson (1961), and McNemar (1962). They argue that statistics apply to numbers rather than to things. The formal properties of measurement should have no effect on the choice of statistics. Recently Gaito (1980) reviewed the entire problem of measurement scales in relation to statistical operations. He observed that in the mathematical statistics literature one does not find scale properties as requirements for the use of various statistical procedures. To Gaito, Stevens's view is based on a confusion between measurement theory and statistical theory. Thus it can be said that although a knowledge of scale types is needed, indices of

similarity such as distances and correlations may be computed even though the presence of interval scales cannot be demonstrated. Most quasi-normal psychological variables can be treated as interval scales. The uncertain reader can peruse Anderson or Gaito to reach a decision.

The Distance Measure

Consider a set of N entities described in terms of k variables, where k is usually much less than the full set of n descriptive variables. The letter k is used in this chapter, and occasionally elsewhere, to refer to a reduced set of factor scores or composite variables that represent k relatively independent measures of descriptive dimensions. A hypothetical data matrix is shown in Table 3-1. Each row represents the score profile of entity i; each column gives the scores received by the N entities on the j variable. Each cell entry represents the value X_{ij} of entity i on variable j.

Suppose two entities a and b are to be compared for similarity. The score profile of a is represented by vector X_a $(X_{a1}, X_{a2}, \ldots, X_{ak})$ and the score profile of b is represented by vector X_b $(X_{b1}, X_{b2}, \ldots, X_{bk})$. Each set of scores may be regarded as coordinates for a vector in k-dimensional space. A *vector* is a quantity that has magnitude and direction. It may be represented by a directed line segment: The length of the vector represents its magnitude, and the direction indicates where the vector is pointing in space. Let the vectors be located in two-dimensional space as shown in Figure 3-1 and assume that the coordinate axes are orthogonal (rectangular). The length of vector a would be the square root of $(X_{a1}^2 + X_{a2}^2)$ and

Table 3-1. Illustrative Data Matrix.

		Variables					
		1	2	. . .	j		k
	1	X_{11}	X_{12}	. . .	X_{1j}	. . .	X_{1k}
	2	X_{21}	X_{22}	. . .	X_{2j}	. . .	X_{2k}
	.	. . .	. . .	. . .	. . .	. . .	. . .
Objects	i	X_{i1}	X_{i2}	. . .	X_{ij}	. . .	X_{ik}
	h	X_{h1}	X_{h2}	. . .	X_{h3}	. . .	X_{hk}
	.	. . .	. . .	. . .	. . .	. . .	. . .
	N	X_{N1}	X_{N2}	. . .	X_{Nj}	. . .	X_{Nk}

Figure 3-1. Effect of Eliminating Scatter on Profile Length and Distance Separation.

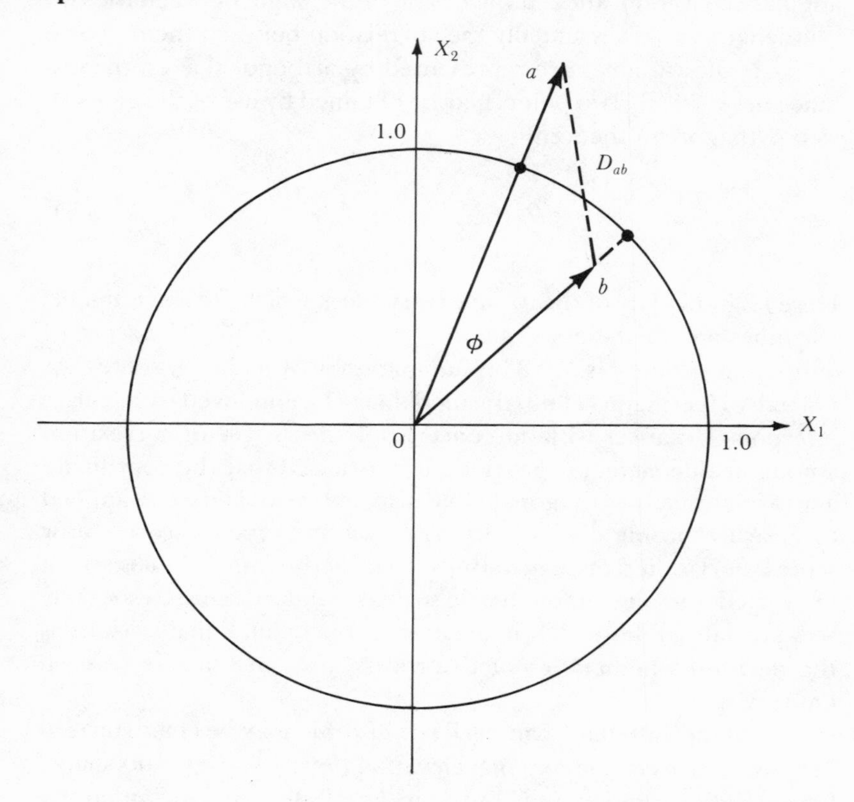

the length of b would be the square root of $(X_{b1}^2 + X_{b2}^2)$. The more similar the profiles of a and b, the closer their vectors; the more dissimilar, the more distant the two vectors. The similarity between the profiles can be evaluated in terms of the distance between the vector endpoints. According to the familiar pythagorean theorem,

$$D_{ab}^2 = (X_{a1} - X_{b1})^2 + (X_{a2} - X_{b2})^2$$

Its square root gives the distance between a and b as shown in Figure 3-1. The figure also reveals that the distance D between vector endpoints is a function of the angle of separation between the profile vectors and their respective lengths. If the vectors are scaled to unit

length, the distance between their endpoints is a function only of the angular separation. Indeed the cosine of the angle between the two unit length vectors is actually the correlation between them.

If the variables are represented by orthogonal axes, the distance between any two points may be obtained by use of the generalized pythagorean theorem:

$$D_{ih}^2 = \sum_{j=1}^{k} (X_{ij} - X_{hj})^2 \tag{3-1}$$

Here j signifies any of the *variates* (variables), which are k in number; i signifies any of the entities $a, b, \ldots, h, \ldots, N$. The score of entity i on variate j is X_{ij}. The summation is over the k variates. As noted by Heermann (1965) the model may be employed to calculate interpoint distances without concern for the degree of correlation among profile elements provided it is assumed that the coordinate axes are mutually orthogonal. The distance formula may be applied to measure profile dissimilarity based on any type of score—raw scores, deviation scores, or ratings. Usually the test variables are all converted into a common metric such as standard scores, especially if the variates differ widely in mean and dispersion. Equally critical is the decision whether or not to process the score profile of each entity.

As indicated in Chapter Two, a profile may be characterized in terms of three components: elevation (level), scatter, and shape. The *elevation* component $\overline{X}_i$ is the mean of all scores for an entity. *Scatter* S_i is defined as the square root of the sum of squares of the entity's deviation scores around its own mean. *Shape* is the information remaining in the score set after removing elevation and equalizing scatter. In a two-dimensional graph the shape may be seen in the rank order of the scores; it shows which are high and which are low relative to the others. Two profiles have the same shape if the ranks of the profile elements are the same. An entity's k raw scores represent k "degrees of freedom," a term used by statisticians to mean freedom to vary. If the mean $\overline{X}_i$ is subtracted out of each raw score, 1 degree of freedom is lost. If the profile deviation scores are divided by the entity's standard deviation in order to standardize it, another degree of freedom is lost.

The effect of removing elevations is illustrated in Figure 3-2. Assume that five entities *a*, *b*, *c*, *d*, and *e* have raw scores on two dimensions X_1 and X_2. Suppose the raw score profiles are (4,−1), (3,1), (2,3), (−1,2), and (1,5). The five means are 1.5, 2.0, 2.5, 0.5, and 3.0, respectively. The deviation score profiles are (2.5, −2.5), (1.0, −1.0), (−0.5, 0.5), (−1.5, 1.5), and (−2.0, 2.0). In Figure 3-2 the points represent the five entities on the two axes. Removing the entity mean from each raw score projects these points onto the line L_1L_2. A point is projected orthogonally to a line by dropping a perpendicular to the line. The deviation scores define L_1L_2 when the coordinates of each entity are plotted. Instead of being located in

Figure 3-2. Effect of Eliminating Elevation from Entity Profile Scores.

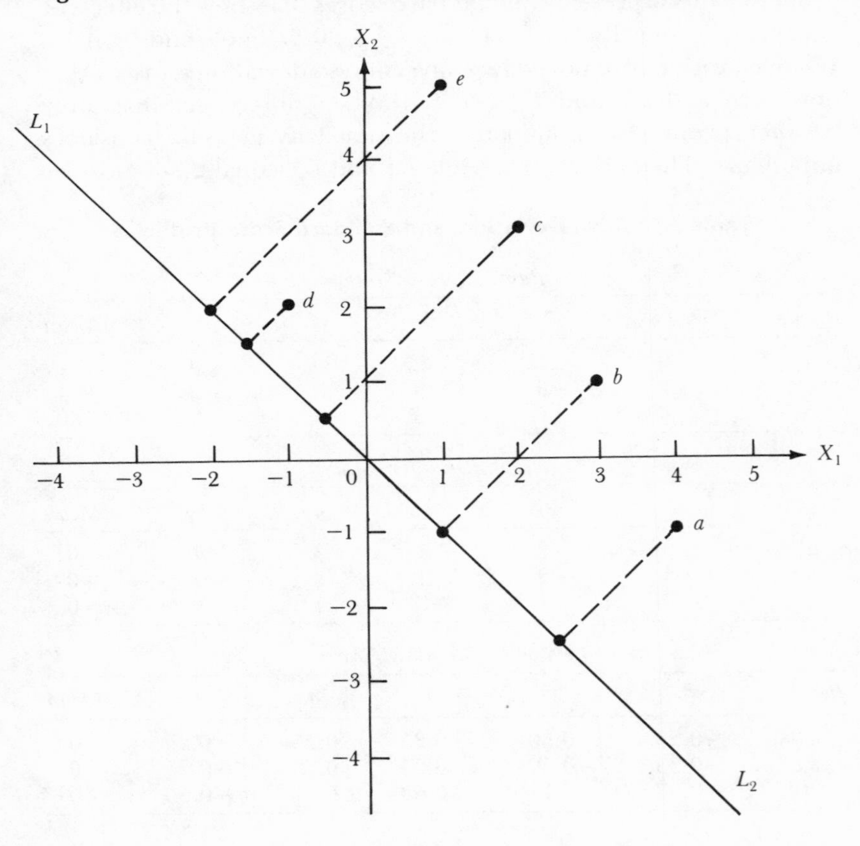

two dimensions, each point is now ordered along one dimension (L_1L_2).

The scatter of scores around the profile means can be represented by the distance from the origin to the endpoint representing the vector. If the scores are standardized within profiles, all vectors are reduced to unit length or distance from the origin. In Figure 3-1 the circle is a unit distance from the origin. Vector *a* becomes shorter and vector *b* becomes longer.

There is a loss of information when level is removed and scatter is equalized; this loss can be illustrated by their effects on the score means and distances between entities. Suppose entities *a*, *b*, and *c* have raw scores on five tests as shown in panel A of Table 3-2. Upon removing elevation the scores and distances are shown in panel B. Panel C presents the standard scores that show the effects of equalizing scatter. By Formula (3-1), D^2_{ab} is 20, D^2_{bc} is 63, and D^2_{ac} is 63. When elevation is removed by conversion to deviation scores, D^2_{ab} is now zero and D^2_{bc} and D^2_{ac} are 58. When differences in scatter between profiles are eliminated, the similarity measure considers only shape. Then D^2_{ab} is zero, while D^2_{bc} and D^2_{ac} equal 2.36.

Table 3-2. Raw, Deviation, and Standard Score Profiles.

Panel A: Raw Scores

Profile	*1*	*2*	*3*	*4*	*5*	*Mean*
a	0	−1	2	4	0	1
b	−2	−3	0	2	−2	−1
c	4	1	−3	1	−3	0

Panel B: Deviation Scores

Profile	*1*	*2*	*3*	*4*	*5*	*Mean*
a	−1	−2	1	3	−1	0
b	−1	−2	1	3	−1	0
c	4	1	−3	1	−3	0

Panel C: Standard Scores

Profile	*1*	*2*	*3*	*4*	*5*	*Mean*
a	−0.25	−0.50	0.25	0.75	−0.25	0
b	−0.25	−0.50	0.25	0.75	−0.25	0
c	0.67	0.17	−0.50	0.17	−0.50	0

Suppose the raw scores in Formula (3-1) are expressed algebraically in terms of deviation scores. Then the distance equation becomes

$$D_{ih}^2 = \Sigma x_{ij}^2 + \Sigma x_{hj}^2 - 2\Sigma x_{ij} x_{hj} + k(\overline{X}_i - \overline{X}_h)^2 \quad (3\text{-}2)$$

where Σx_{ij} and Σx_{hj} represent deviation scores of the two entities and $\Sigma x_{ij} x_{hj}$ denotes the covariance. Now if we substitute equivalents for the first three terms on the right, distance is seen as

$$D_{ih}^2 = S_i^2 + S_h^2 - 2Q_{ih} S_i S_h + k(\overline{X}_i - \overline{X}_h)^2 \quad (3\text{-}3)$$

where Q_{ih} represents the correlation between the two entities. If $(S_i - S_h)^2 + 2S_i S_h$ is substituted for the first two terms and the equation is simplified, then

$$D_{ih}^2 = k(\overline{X}_i - \overline{X}_h)^2 + (S_i - S_h)^2 - 2S_i S_h(1 - Q_{ih}) \quad (3\text{-}4)$$

The terms on the right represent differences in elevation, differences in scatter, and differences in shape weighted by scatter.

Formula (3-4) indicates that the generalized distance concept includes separate components of elevation, scatter, and shape. Should the scores be standardized and elevation eliminated, the distance measure becomes

$$D_{ih}^2 = 2(1 - Q_{ih}) \quad (3\text{-}5)$$

This expression indicates that correlations between profiles are simply measures of distance in $k - 2$ space where scatter and elevation have been eliminated.

Elevation can be shown to correspond fairly closely to any general factor underlying the set of tests. Thus subtracting the entity mean out of each raw score loses one dimension. The scatter of scores around the mean of a profile is represented by the length of the profile vector. If the profile is jagged, the scatter is high and the vector is long. If the profile is relatively flat, the scatter is low and the vector is short. If the scores are standardized, all the vectors will be of unit length.

The characteristic shape of a profile may be expressed geo-

metrically by the direction of the vector in variable space. Consider, for example, two entity vectors each of unit length in two-dimensional space as in Figure 3-1. If entity *a* has a high score on variable X_1 and a zero score on variable X_2, its vector will coincide with X_1. If entity *b* has a high score on X_2 and a zero score on X_1, its vector will coincide with X_2. The two vectors will then be separated by an angle of 90°. As the rank order and size of the two sets of scores become similar, the angle between their score vectors becomes acute. When their scores coincide, the vectors will coincide and the profile orientations will be the same. Thus the vector's orientation is determined by its high and low scores.

Metric Measures

The most mathematically advanced of the distance functions are those called *metrics.* A metric model makes the assumption that the scale of measurement is an interval scale. A nonmetric model requires only the assumption that the scale of measurement is an ordinal scale.

A distance function $D(x, y)$ of a pair of points of a set E is a metric if it satisfies the following conditions:

1. $D(x, y) \geq 0$. The distance between two points is never negative.
2. $D(x, y) = 0$ if and only if $x = y$. The distance between two identical points is always zero; the distance between two different points is never zero.
3. $D(x, y) = D(y, x)$. Distance is symmetric; that is, the distance between x and y is the same as that between y and x.
4. $D(x, y) + D(y, z) \geq D(x, z)$. The sum of the distances between two points (x and z) and a third point (y) is always greater than or equal to the distance between these two points. This property is known as the *triangular inequality:* The length of one side of a triangle must be no longer than the sum of the lengths of the other two sides. A function that satisfies the first three axioms is called a *semimetric;* a *metric* satisfies all four conditions; an *ultrametric* (Johnson, 1967) satisfies a fifth condition.
5. $D(x, z) \leq \max[D(x, y), D(z, y)]$. The ultrametric inequality axiom ensures that the pair function is monotonically increasing or

decreasing depending on whether it is a similarity or a dissimilarity function. The axiom requires that such triples of distances must form either an equilateral or an isosceles triangle where the base is shorter than the two equal sides. The importance of this requirement will become evident when we discuss hierarchical clustering algorithms.

Other Distance Measures

The most familiar distance measure is the euclidean metric. The most general distance, however, is known as a *Minkowski metric.* It includes as special cases euclidean and other metrics. The distance is defined as

$$D_{ih}^2 = \left[\sum^{k} |X_{ij} - X_{hj}|^p \right]^{1/p} \tag{3-6}$$

By choosing various values of p one can obtain a variety of different metric distance functions. If $p = 1$, the *city-block* or *taxicab* metric is obtained:

$$D_{ih} = \sum^{k} |X_{ij} - X_{hj}| \tag{3-7}$$

This distance function represents the sum of the absolute (positive) values of the difference between entities for each profile element. Another version, called the "mean character difference," divides the sum by k, the number of variables. One justification given for the city-block metric is that when two entities are described by two variates whose scale units are of equal value, they should be the same distance apart—whether 2 units apart on each variable or 1 unit apart on one variable and 3 units apart on the other. The method is used in several cluster algorithms.

In the euclidean distance, $p = 2$. In order to make the distance function comparable in magnitude across studies, it is becoming increasingly common to divide the squared differences by k, the number of variables.

Another index r_p, proposed by Cattell (1949), includes all the

information obtained in D^2 but ranges from + 1 to − 1. The change is achieved by means of a transformation:

$$r_p = \frac{2K - \Sigma d^2}{2K + \Sigma d^2} \qquad (3\text{-}8)$$

where $2K$ equals twice the median chi square corresponding to the given number of variates and Σd^2 is simply D^2. Although the intent was to create an index comparable to a correlation without loss of information, there is no reason why there should be a limit on dissimilarity between two entities.

In using D as the definition of similarity, or closeness of profiles, the problem often arises as to whether it is appropriate if the variables defining the profiles are correlated. Some investigators, such as Overall (1964), argue for use of a generalized distance measure D_M^2 developed by Mahalanobis (Rao, 1952).

The D_M distance is a measure of similarity in which all the original variates have been transformed into uncorrelated orthogonal components:

$$D_M^2 = d'_{hi}\mathbf{W}^{-1}d_{ih} \qquad (3\text{-}9)$$

where $\mathbf{W}^{-1}$ is the inverse of the pooled within-group variance-covariance matrix and d_{ih} is a column vector of score differences between entities i and h. The d'_{hi} (row vector) is the transpose of d_{ih}. Rather than being put off by this forbidding formula, remember that it is meaningful and can be calculated by computer. The Mahalanobis measure was developed to determine the distances between groups but has been adapted to differences between individuals. When it is applied to individual entities, the intercorrelations of the variates for a specific reference group should be used. The major limitation of this measure is that it weighs reliable and important elements equally with the unreliable and unimportant ones. For variates that are standardized and uncorrelated, D_M is the same as D. In a study reported by Hartigan (1975) the Mahalanobis distance was found to reduce the clarity of clusters; within-cluster distance increased but between-cluster distances did not.

Correlation Coefficients

The Pearson product-moment correlation Q was first used to assess the resemblance between two persons by Stephenson (1936). Since then it has been widely used as a measure of similarity in profile shape. The correlation between entities i and h may be defined as

$$Q_{ih} = \frac{\Sigma(X_{ij} - \overline{X}_i)(X_{hj} - \overline{X}_h)}{\sqrt{\Sigma(X_{ij} - \overline{X}_i)^2}\sqrt{\Sigma(X_{hj} - \overline{X}_h)^2}} \tag{3-10}$$

where $\overline{X}_i$ and $\overline{X}_h$ are the entity means and the denominator terms represent the scatter. Since elevation is removed and scatter is equalized, the correlation coefficient measures similarity in $k - 2$ space. The range of coefficients is from $+1$ to -1.

Because the variables may differ markedly in metric it is critical to standardize each measure to have a mean of zero and a variance of 1 prior to correlating two entities. Furthermore the index will be most meaningful when the variables are relatively uncorrelated. For this reason it is a common practice to intercorrelate the n variables and reduce them to a smaller set of k measures by means of a dimensional analysis to be described later.

Cohen's Coefficient r_c. Cohen (1969) observed that the product-moment correlation r (also called Q), as a measure of profile similarity, suffers from the defect that its value depends on the direction in which variables are measured. Consider a profile of five variables rated on five 7-point rating scales of Sociable versus Detached, Agreeable versus Hostile, Independent versus Conforming, Dominant versus Submissive, and Liberal versus Conservative. Any profile element may be reflected or reversed with no change in its differential measurement. A person can score high on Sociable or on Detached, high on Agreeable or on Hostile, and high on Dominant or Submissive. Since the direction of measurement is arbitrary, none of these reflections should change the substantive conclusion from an analysis of data. If the similarity of two profiles is evaluated by Q over k variables, however, the correlations may change markedly with changes in direction (Tellegen, 1965).

A new profile similarity coefficient proposed by Cohen (1969) is meant to be invariant over reflection about m, the neutral point, of

any number of variables. The procedure involves addition of k new scales, each a reflection of the original scales. Cohen reasons that since the original directions are arbitrary, each element is represented both in its original direction and its opposite. Then for the $2k$ elements a product-moment correlation is found. The new coefficient is defined as

$$r_c = \frac{\Sigma(X_{ij} - m)(X_{hj} - m)}{\sqrt{\Sigma(X_{ij} - m)^2}\ \sqrt{\Sigma(X_{hj} - m)^2}} \tag{3-11}$$

A scale value is reflected by placing it symmetrically on the opposite side of the neutral point. The neutral point, which is the psychological center, can be defined by the scale or by standardization means. If the scale value of the neutral point is m, the reflected score $X = 2m - X$. Suppose, for example, that entity A's ratings are 1, 7, 5, 3, 4 and B's ratings are 3, 6, 5, 1, 4. Let us now apply r_c to the k elements and their k reflections. Then A's extended scores are 1, 7, 5, 3, 4, 7, 1, 3, 5, 4; B's extended scores become 3, 6, 5, 1, 4, 5, 2, 3, 7, 4. The value of r for k elements is 0.75, while the value of r_c becomes 0.919 for the $2k$ elements.

When m represents the actual means of X_{ij} and X_{hj}, then r_c becomes the ordinary product-moment correlation Q_{ih}. When $m_i = m_h = 0$, the formula becomes what is known as the *congruency coefficient.*

The Congruency Coefficient. The congruency coefficient C is frequently used to assess similarity between profiles. It may be defined as the sum of the normalized cross-products of two profiles:

$$C = \frac{\Sigma X_{ij} X_{hj}}{[\Sigma X_{ij}^2\ \Sigma X_{hj}^2]^{1/2}} \tag{3-12}$$

Since the entity means are not subtracted out in either the numerator or denominator, information regarding elevation is retained. Scatter, on the other hand, is equalized because the operation normalizes both profile vectors—that is, it transforms them to unit length.

Suppose we use the five ratings previously listed for A and B to compute the value of the congruency coefficient. The sum of cross-

products in the numerator is 89 and the denominator of C equals 93.2. Thus the value of C is 0.95 whereas Q is 0.755.

Coefficient C was developed independently by Burt (1937), Tryon (1955), Tucker (see Harman, 1976), and Sjoberg and Holley (1967). Holley and Guilford (1964) proposed another index called G that is a special case of r_c when the elements are scored dichotomously. Index G is described later in the chapter in the section concerned with measuring association between binary variables.

Intraclass Correlations. The intraclass correlation represents another alternative to distance measures and to product-moment correlations. Like the congruency coefficient, the intraclass r_{in} is sensitive to similarities in profile shape as well as to differences in profile elevation. A value of 1.0 can be obtained only if two profiles have the same scores on all profile elements. The intraclass correlation is based on a one-way analysis of variance. For a more extensive discussion see Haggard (1958) and Bartko (1976). The formula for r_{in} in terms of mean squares is MSB − MSW/MSB + MSW. In other words, r_{in} is the ratio of mean squares between variables minus mean squares within variables over total mean squares. Computational formulas for the sums of squares are as follows:

$$B = \frac{\sum^{k}\left(\sum^{2} X\right)^2}{2} - \frac{\left(\sum^{2}\sum^{k} X\right)^2}{2k} \quad \text{(Between SS)} \tag{3-13}$$

$$T = \sum^{2}\sum^{k} X^2 - \frac{\left(\sum^{2}\sum^{k} X\right)^2}{2k} \quad \text{(Total SS)}$$

$$W = T - B \quad \text{(Within SS)}$$

Let k denote the number of profile elements (variables) and p the number of profiles. The degrees of freedom for calculating the mean squares are then $k - 1$ for B, $kp - 1$ for T, and $kp - k$ for W. A useful computational example may be found in Edelbrock and McLaughlin (1980).

Edelbrock (1979) conducted a series of cluster analyses designed to check accuracy of recovery of known clusters in a mixture model. Use of intraclass r_{in} rather than distance in these cluster analyses yielded the most accurate recovery rates. Edelbrock and

Achenbach (1980) utilized r_{in} as a measure of profile similarity in the process of developing a typology of child behavior patterns. Their comparison with Q correlation indicated that r_{in} resulted in greater differentiation among profile types—perhaps because r_{in} included information regarding elevation whereas Q did not. A more critical view of r_{in} is taken by Cronbach and Gleser (1953). Using a simple two-variable illustration, they show that even when D is the square root of 0.02, r_{in} is -1.0. It is thus evident that the index needs more critical examination.

Rank Correlation Coefficients. Indices of similarity are sometimes needed for scores ranked from highest to lowest within the profile or for objects ranked from highest to lowest. Judges are sometimes asked to rank a set of literary essays, for example, and subjects may be asked to rank 10 values in order of importance to them. The Spearman rank correlation coefficient, rho, the first such measure to be developed, may be expressed as

$$r_s = 1 - \frac{6\Sigma d^2}{k^3 - k} \tag{3-14}$$

where X_{ij} is the rank given by entity i to variable j and where $d_{ij}^2 = \Sigma_j(X_{ij} - X_{hj})^2$ represents the differences between the ranks by entity i and entity h to the k elements. When tied scores occur, each variable is assigned the average of the ranks that would have been assigned if no ties had occurred. Suppose A ranked five essays, 1, 3, 4, 2, 5 while B ranked them 2, 1, 3, 5, 4. Application of Formula (3-14) yields a rank correlation of 0.20.

Kendall's (1948) coefficient tau is equally suitable as a measure of rank correlation. It is based on the direction of difference between all possible variables. This is defined as follows:

$$\text{Tau} = \frac{\text{actual score}}{\text{maximum possible score}} = \frac{S}{\frac{1}{2}k(k-1)} \tag{3-15}$$

Tau is a function of the minimum number of inversions or interchanges between k neighbors required to transform one ranking into another. The actual score S is computed on the basis of the number of ranks in natural order minus the number of rank inversions. The

maximum possible score is estimated from the combination of k things taken two at a time or $k(k-1)/2$.

Tau and rho usually differ in numerical value when both are computed from the same pair of rankings. Generally tau is smaller in value. The two have different underlying scales and, therefore, are not directly comparable. Both, however, have identical power to reject the null hypothesis. Discussions of both indices may be found in Siegel (1956). Computer programs are also available.

Measuring Association

There are numerous measures of association for binary variables in the literature. These measures are also known as *matching coefficients.* It is questionable whether a long list of these measures would be of much value, however, since many of them are used less frequently in social science than in biology. Measures of association are pair functions that measure agreement between two entities over n two-valued variables. Association measures take values in the range from 0 to 1. The characteristics are usually coded 1 or 0 indicating the presence or absence of a characteristic. The frequency of co-occurrence or agreement and the frequency of disagreement are summarized in Table 3-3. In this contingency table, category a represents the number of agreements or the joint presence of both characteristics. Category d represents the absence of both characteristics. Categories b and c indicate the number of disagreements or mismatches.

The various coefficients that have been proposed differ mainly on whether or not to include negative matches and mismatches. Another source of difference concerns the weight to assign

Table 3-3. 2 × 2 Contingency Table.

		Entity i		
		0	1	Total
Entity h	1	b	a	$a+b$
	0	d	c	$c+d$
	Total	$b+d$	$a+c$	n

the matched and unmatched pairs. There are no rigid rules regarding inclusion or exclusion of negative matches. Especially troublesome is the question of what to do about the 0-0 cell or d if it represents absence of both characteristics. If d represents a high proportion of the total, it would be wise to exclude it. The most popular indices are given in Table 3.4.

The first measure is the probability that the two entities both score 1 on a randomly selected variable or that both manifest the same profile elements. The second coefficient, called the *similarity,* is ascribed to Jaccard. It represents the probability that both entities score 1 on a randomly chosen variable given that all 0-0 matches are excluded. The third measure is the probability that the two entities have the same score (either 1 or 0) on a randomly chosen variable. The complement of the third coefficient is $(b + c)/n$, a distance measure derived from D^2 by using binary values of 1 or 0 in the formula. This disagreement index is not a matching coefficient since it does not vary between 0 and 1. The fourth coefficient, known as Dice's coefficient, is monotonic with Jaccard's index. It gives more weight to matches than to mismatches.

Consider two entities A and B rated on the presence or absence of 10 binary characteristics shown in Table 3-5. If a count is made on matches and mismatches, the corresponding four cells of the contingency table (Table 3-3) will be 3 for a, 2 for b, 1 for c, and 4 for d. If these agreements and disagreements are substituted in the matching coefficients of Table 3-5, the values are 0.30, 0.50, 0.70, and 0.66. This example illustrates the effects of weights and the exclusion of 0-0 matches.

Table 3-4. Matching Coefficients.

1. $\dfrac{a}{a + b + c + d}$

2. $\dfrac{a}{a + b + c}$

3. $\dfrac{a + d}{a + b + c + d}$

4. $\dfrac{2a}{2a + b + c}$

Table 3-5. Ratings for Two Entities on 10 Categories.

		Categories									
		1	2	3	4	5	6	7	8	9	10
Entity	*A*	0	1	1	0	0	1	0	0	0	1
	B	0	1	0	0	1	1	0	0	1	1

Holley and Guilford (1964) have proposed G as an index of profile similarity for dichotomous data. Three expressions of the index are

$$G = p_c - p_n = (a + d) - (b + c) = 2p_c - 1 \qquad (3\text{-}16)$$

where p_c is the proportion of agreeing responses (yes-yes or no-no) and p_n is the proportion of nonagreeing responses. Application of G to any 2×2 table in effect replaces a and d each by their average and replaces b and c each by their average. It also equates means and variances in the variables correlated.

Index G is invariant over variable reflection for the same reasons given for Cohen's index r_c and for index H. It is to be preferred to a Pearson correlation (phi coefficient) since the value of phi is influenced by the marginal totals. Phi and G are identical when p and q (marginal totals) equal 0.5. Index G is also to be preferred for a factor analysis of entity correlations. The first factor will be smaller and will therefore present fewer difficulties for finding interpretable solutions. For the preceding example,

$$2p_c - 1 = 2(\tfrac{7}{10}) - 1 = 0.40$$

Choosing a Similarity Index

A variety of indices have now been described. The question remains, however, whether differences in elevation or differences in scatter should be taken into account. Cronbach and Gleser (1953), Cattell (1949), and Fleiss and Zubin (1969) have argued for the use of the distance measure. In general they judge it undesirable to eliminate elevation. In certain studies two objects should not be

regarded as similar if their profiles have the same shape but they differ in elevation. In the MMPI, for example, elevation of scores above 70 on scales Hs, D, and Hy means something different from the same pattern at 60. In measuring overall ability, moreover, the pattern of test scores is qualitatively different in the dull individual and the bright one.

On the other hand, there are studies in which the elevation component is of no interest. For example, the scores of a well-constructed personality inventory are not likely to involve a general factor. Usually it is possible to reflect all the scales in such a way that all the intercorrelations are positive or close to zero. In such cases the shape of the profile should be emphasized. In psychopathology, for example, the diagnostic categories, with but few exceptions, are based primarily on the profile's shape.

An alternative strategy to the use of the distance measure is to differentiate among elevation, scatter, and shape successively. The researcher can then determine the independent contribution of each source of information to cluster differentiation, to hierarchical arrangement, or to the prediction of external criteria. There is no question that the distance measure tends to confound the three kinds of information: D^2 can represent a large difference between two profiles on one dimension or the sum of many small differences on all the variables. The distance measure, moreover, will evaluate as "close" a subset of flat profiles to which shape indices are insensitive. In a cluster analysis it helps to bring together subjects who manifest various response biases, such as acquiescence to authoritarian statements or socially desirable items.

Although the correlation coefficient has been widely used to measure profile similarity, it has also been criticized. If profiles are parallel the correlation is unity, but the converse is not true: Two profiles may correlate +1 even if they are not parallel to each other. They need only to be monotonic functions (uniformly increasing or decreasing) of each other (Fleiss and Zubin, 1969). Another limitation in the use of a correlation to measure profile similarity is that its value depends on the direction in which variables are measured. This problem was discussed in connection with Cohen's r_c. On the other hand, many cluster analysts prefer clusters that reflect profile shape differences rather than elevation. (In a later chapter it is shown

that known cluster structure can be recovered with equal accuracy with correlating distance measures.) In judging members of a type it is the rank order of relevant profile components that distinguishes one type from another.

A way out of the choice between correlations and distance functions is to assess separately all three components of shape, level, and scatter. Guertin (1966) suggests that correlation should first be used to form homogeneous groups based on profile shape; then each shape cluster can be subdivided on the basis of elevation and scatter. Skinner (1978) too suggests placing profiles into relatively homogeneous subgroups on the basis of shape, followed by differentiation within each group on the basis of scatter and elevation. An example of such use may be found in Skinner and Jackson's (1978) study of T-score profiles on the MMPI. They identified three relatively homogeneous psychiatric subgroups on the basis of profile shape. The groups were then differentiated on the basis of scatter and elevation.

Summary

In this chapter I have reviewed the problems relating to types of variables, scales of measurement, and indices of similarity. Definitions and illustrations were given for three kinds of variables: continuous, discrete, and unordered polytomies. The terms entity, attribute, and quality were defined and illustrated. The four scale types, nominal, ordinal, interval, and ratio were distinguished and the permissible transformations for each were listed. Two viewpoints regarding the statistical requirements for distance functions and correlations were sketched and compared. A generalized distance measure frequently used to assess the similarity between entity profile was delineated. Each entity was described in terms of a profile which has three components: elevation, scatter, and shape. The effect of removing elevation or scatter on the distance function was shown algebraically and illustrated in figures and tables. Correlation coefficients as measures of angular separation between profile vectors were related to the distance function. The coefficients described, discussed, and compared included the product moment correlation Q, Cohen's r_c, the congruency coefficient C, the intraclass

r_{in}, the rank order coefficient r_s, and Kendall's tau. Four matching coefficients for measuring agreement between binary variables were also presented and illustrated. The chapter concluded with a discussion as to whether researchers should use a distance function that takes elevation and scatter into account or whether they should use correlation coefficients which do not.

In the next chapter we turn to a consideration of ordination —the process of constructing a low-dimensional mapping of the entities in attribute or entity space. In this way the researcher can visualize the location and possible separation of swarms of points that represent clusters.

CHAPTER 4

Methods of Ordination

This chapter outlines some of the methods used to map individual entities in attribute spaces with very few dimensions. The process, called *ordination,* resembles the procedure used to locate major cities in the United States in a geographical atlas. Since the space is two-dimensional, each city is located by an east-west coordinate and a north-south coordinate. Chicago, for example, might be located on a map at E-4. Several methods of ordination are described and compared here. Then attribute space is differentiated from entity space. The value of ordination lies in its power to provide the researcher with a visual picture of the various entities and the patterns they might exhibit. The coordinates of this low-dimensional space are also useful as a basis for cluster analysis.

Meaning of Ordination

The cluster analyst will at times find it useful to obtain a low-dimensional mapping of a set of objects under study. The space in which the N points are embedded can then be examined visually to discover any grouping. Objects close together in this space can then be segregated into subsets by visual inspection. The term *ordination* has been used to refer to this process of locating the position of each

entity in a space of small dimensionality. The word originated in biology (Goodall, 1953), and the process has been widely applied by numerical taxonomists (Sneath and Sokal, 1973). The most popular methods of ordination are principal-component analysis, principal-coordinate analysis (Gower, 1966), and multidimensional scaling (Shepard, 1962; Kruskal, J. B., 1964).

Before describing these procedures it is important to point out that ordination may be applied to locate entities in *attribute space* or in *entity space.* Consider the data matrix of N rows (entities) by n columns (attributes). A-space (attribute space) has formally n dimensions, one for each attribute; in A-space there are N points representing the entities. E-space (entity space) has N dimensions, one for each entity; in E-space there are n points representing the attributes. Ordination in A-space customarily begins with an $n \times n$ matrix **R** of correlations among the attribute pairs across the sample of entities. A dimensional analysis of the **R** matrix (called an R analysis) yields the coordinates of the n attributes on r coordinate axes. Computation of the scores of the N entities on r coordinate axes permits a mapping of the entities in A-space. Ordination in E-space begins with an $N \times N$ matrix **Q** of correlations among the entity pairs across the n attributes. A dimensional analysis of **Q** (Q analysis) yields the coordinates of the N entities on r entity coordinate axes. Most researchers then map the N entities on the first two or three entity axes.

The R analysis is more direct and more economical than Q analysis, especially if there are more entities than attributes. Moreover, the entity groupings can be understood better when embedded in the more meaningful attribute space of R analysis. Only when there are more variables than entities in the data matrix is it more useful to conduct a Q analysis. Suppose, for example, that an intelligence test of 300 items is completed by 100 persons. An R analysis of the 300 × 300-item matrix would be much more costly and difficult than a Q analysis of the 100 × 100 matrix of correlations between persons.

With respect to dimensions, Cattell (1978, p. 326) observes: "Q analysis tells us nothing we do not know from R technique, and vice versa, with a small exception. As Horst (1965, p. 324) says, "It does not seem to be generally recognized that for a given set of

metricizing operations . . . it does not matter whether we factor one set of categories [attributes] or the other [entities].'' The conditions under which these reciprocal relationships hold will be explained later, but the position taken here is that the differences between the two approaches are minor. In most cases there will be far more entities than profile elements. Since an *R* analysis is more direct and economical than a *Q* analysis, it is to be preferred.

Dimensional Analysis

In Chapter Two, the aims of dimensional analysis were said to be data reduction, the generation of hypotheses regarding underlying structure, and the testing of hypotheses regarding factor structure. Before elaborating any further, however, it is necessary to differentiate between two models of dimensional analysis: principal-component analysis and factor analysis.

In the principal-component or maximum-variance model the aim is to account for as much of the variance as possible in the successive components. The model may be expressed as

$$Z_j = a_{j1}F_1 + a_{j2}F_2 + \cdots + a_{jn}F_n \qquad (j = 1, 2, \ldots, n) \qquad (4\text{-}1)$$

where Z_j is the observed standard score on test j, the a_j's represent the component coefficients or loadings, and the F's are the component scores. In other words, the observed score on variable j is expressed as a linear combination of weighted component scores. When this model is adopted, the procedure is known as *principal-component analysis* (PCA). Two restrictions are built into the procedure:

1. The first component, and each successive one, must account for the maximum variance possible.
2. The second principal component is the linear combination of the observed variables that accounts for as much of the remaining variance as possible subject to the second restriction—that it be uncorrelated with the previous one.

That process is repeated until all the variance is accounted for by n components. Since the last few components are very small and

statistically unreliable, they are usually discarded. The components may be summarized in an $n \times r$ matrix **A** of component coefficients. The entries represent the correlations of each variable with each of the r components. Following this, r component scores can be computed for each entity. The component scores are obtained by finding for each entity the numerical value of the linear combination. The scores can be summarized in an $r \times N$ matrix **F**. Expressed in matrix notation $\mathbf{Z} = \mathbf{AF}$, where **Z** represents the observed scores, **A** the component coefficients, and **F** the component scores.

The second model, called *factor analysis*, is designed to reproduce as accurately as possible the correlations among the observed variables. The aim is to express each observed variable in terms of a linear combination of hypothetical factors or constructs. The model, stated explicitly for the value of variable j for entity i, is

$$Z_{ji} = a_{j1}F_{1i} + a_{j2}F_{2i} + \cdot \cdot \cdot + a_{jr}F_{ri} + d_jU_{ji} \qquad (j = 1, 2, \ldots, n) \qquad (4\text{-}2)$$

where each of the n variables Z_{ji} is described in terms of r common factors and the unique factor U_j. Here A_{jp} represent the *factor loadings* or coefficients and F_{pi} represent the common-factor value for entity i. Each variable is hypothesized to derive from three sources of variation. The part that a variable shares with other variables is called the *common factor* variance. The specific aspects of a variable not shared by other variables is called the *specific factor* variance. The third source of variation, random error or unreliability, is referred to as the *error* variance. The part that is unique to a given variable, the combined specific and error parts, constitutes the *unique factor* variance. Since each observed variable is describable in terms of r common factors and n unique factors, the aim is to find the smaller set of common factors and to remove the unique variance. As with principal components, the factors are summarized in an $n \times r$ matrix **A** of factor coefficients. Again the factor loadings represent the correlations of attribute variables with the r common factors. The factor scores must be estimated, however, rather than computed like the component scores. The factor scores are recorded in an $r \times N$ matrix **F**.

Before proceeding, let us review the distinction between principal components and common factors. Components simply

summarize the data in the form of new dependent variables. Common factors represent hypothetical independent variables that explain the correlations or covariances between variables in terms of processes that generated the observed data. The factor model thus assumes that the observed variables are linear combinations of one or two hypothetical variables.

Principal-Component Analysis

The process of principal-component analysis can be illustrated geometrically. Consider the ordinary scatter diagram for the correlation between two variables: One variable is measured on the X_1 axis, the other on the X_2 axis at a right angle to the first. Each entity is represented by a point on the diagram. If the variables are distributed normally, the contours of equal density will be ellipses or, in the case of three or more dimensions, ellipsoids. Suppose we have measures for N entities on two variables or attributes (X_1,X_2). The raw scores or measures are plotted in Figure 4-1*a* and portrayed as an elliptical swarm. A common metric is needed to make the measures comparable because they may differ (height in inches, for instance, and weight in pounds). To achieve this metric the distribu-

Figure 4-1. Principal Axes for Bivariate Distribution.

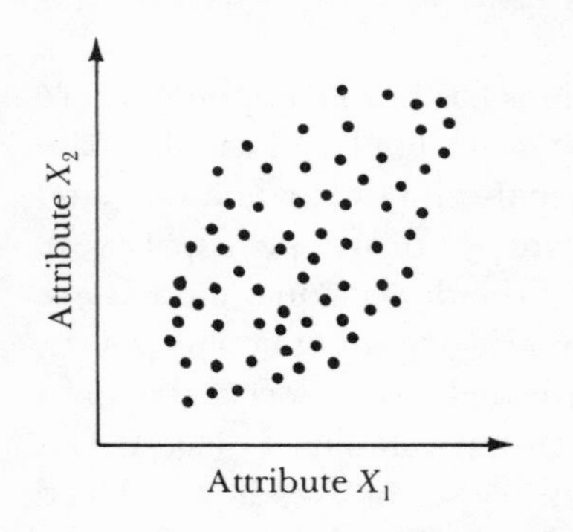

(a) Scatter diagram of raw scores on two variables.

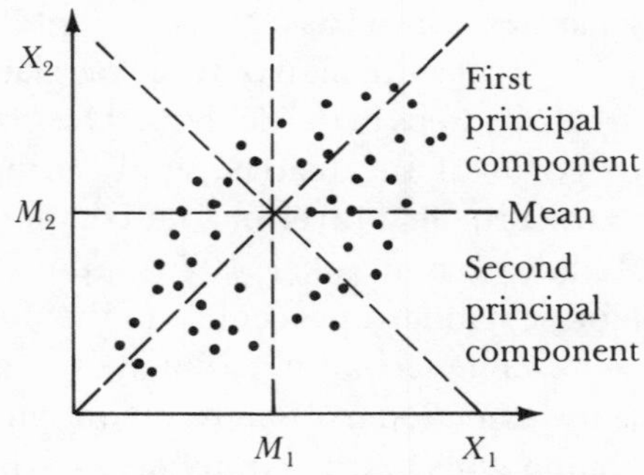

(b) Scatter diagram of standardized scores on two variables.

tions are standardized to zero mean and unit dispersions. Each entity is then referred to a new set of axes corresponding to the means of each variable as in Figure 4-1*b*. Thus each entity now has a set of new coordinates in standard-score form.

To find the maximum separation of entities in one dimension, the points must be projected onto a line that lies between the two axes. This line of best fit is one for which the sum of the perpendiculars to the line from the pairs of coordinates is a minimum. The dispersion or variance of pairs is then at a maximum. This line of maximum variance is known as the principal axis of the ellipse (or the axis of the ellipsoid if three variables are involved). The variance of the coordinates is known as the *eigenvalue.* The axis set at right angles to the first is the minor axis of the ellipse or the second principal component. In terms of analytic geometry, the process of finding the lines of best fit involves a rotation of the coordinate axes to a new frame of reference in the total variate space (Figure 4-1*b*).

In the case of three variables the principal axis joins the center of the ellipsoid to its surface and ends perpendicular to the tangent at the point where the axis meets the surface. A football is a typical three-dimensional ellipsoid. It has three axes, one for each dimension: The first is the long axis joining its two ends; the other two are equal in length and at right angles to the first axis. The *eigenvalue* (latent root) specifies the length of the axis, while the *eigenvector* (latent vector) specifies its direction. The variance of a component is the square of the distance from the center of the ellipse to its circumference. The procedure thus *decomposes* the variance and separates it into basic parts.

Once the matrix **R** of correlations has been computed, there are three steps in the analysis. The first step is to place 1's in the main diagonal of the matrix, apply principal-component analysis, and extract the first three or four components (known as principal axes). As Sneath and Sokal (1973) report, three-dimensional models are almost standard procedure in the numerical taxonomic literature.

The second step, not always followed, is to select and apply a transformation to obtain a unique and interpretable set of axes. The components extracted define an arbitrary set of axes in a reduced dimensional space in which each variable is a point. The problem of

rotation is to select unique axes to describe the space. To provide a meaningful solution, exploratory studies usually follow a principle called *simple structure*—a factor structure characterized by parsimony and simplicity. When simple structure is present, each attribute variable is describable in terms of either one or two components. Each component has significant correlations with only a limited set of variables and near-zero correlations with all others.

The third step is to compute the component score matrix **F**. The values in this matrix are the coordinates of each entity in the low-dimensional space. The component scores may be calculated directly since estimation is not involved. To solve for **F**, use is made of factor matrix **A**, the matrix of observed scores **Z**, and the diagonal matrix $\mathbf{\Lambda}$ of the r eigenvalues. The formula may be found in Harman (1976) and elsewhere. Most general-purpose computer packages—SPSS, BMDP, OSIRIS, and SAS, for example—contain programs for computing component scores.

The final step is to plot the values in each row of matrix **F** against the values of every other row. The r rows represent the principal components and the N columns represent the entities. The plots indicate the position of each entity relative to the two axes. They should be scrutinized for clusters of points that separate from other clusters. If the points appear randomly placed, no clusters are evident and the sample may be homogeneous. If the findings look promising, the entity score profiles can be cluster-analyzed by some selected algorithm. Chapter Eight presents a table of correlations among 15 body types and their analysis into three dimensions.

Some investigators have found that component scores do indeed have advantages over unmodified sets of standard scores. Burket (1964) showed that the most accurate predictions obtainable, accomplished independently of a criterion, are those based on the largest principal components. Likewise Morris (1980) has shown that reduced-rank principal components are significantly more accurate upon cross-validation than the conventional full-rank composite scores. By the *rank* of a matrix is meant the number of linearly independent columns or rows. Thus the first few principal components provide a sound basis for later cluster analysis as well as for ordination.

Common-Factor Analysis

If the common-factor model is chosen, the analysis begins as in principal-component analysis with the correlation matrix. Some estimate is needed of the amount of variance each variable has in common with the remaining variables. This estimate h_i^2, called the *communality,* is placed in the diagonal cells of the correlation matrix. Usually the square of the multiple correlation coefficient of each variable with all others is selected as the estimate. If the same algorithm is applied as in extracting principal components, the method is called a *principal-axis analysis.* Other methods of factor extraction, such as maximum likelihood and minimum residuals (minres), are also available. Useful outlines may be found in Harman (1976) and Gorsuch (1974).

Should only a low-dimensional solution be sought, there is no problem in deciding on the number of common factors to retain. Otherwise some rule for determining the number of factors must be followed. The common factors retained are then rotated in accordance with some analytic (mathematical) procedure such as varimax that yields simple structure solutions. Unlike principal components, common-factor scores must be estimated. Components are linear composites of observed variables and hence have precise score values. Common factors and the accompanying unique factors cannot be so expressed; they can only be estimated. Green (1976) has a good discussion of the problem. Procedures for estimating factor scores are described by Kaiser (1962a) and Harman (1976). Familiar computer packages that produce factor scores are SPSS (Nie and others, 1975) and BMDP (Dixon, 1981).

Principal-Coordinate Analysis

Gower (1966) has developed a procedure called principal-coordinate analysis. The analysis begins with a matrix of interentity correlations Q_{ij} that are related to interentity squared distances as follows:

$$d_{ij}^2 = 2(1 - Q_{ij}) \tag{4-3}$$

or

$$Q_{ij} = 1 - \tfrac{1}{2}d_{ij}^2$$

To apply the Gower procedure:

1. Transform the correlation matrix entries into squared distance by Formula (4-3).
2. Form the matrix **E** with elements $e_{ih} = -\tfrac{1}{2}d_{ih}^2$.
3. Compute the $N \times N$ matrix **F** by double-centering **E**. Its elements consist of $f_{ih} = e_{ih} - \bar{e}_i - \bar{e}_h + \bar{e}$, where $\bar{e}_i$ denotes the row mean, $\bar{e}_h$ the column mean, and $\bar{e}$ the grand mean. In other words: Subtract from each e_{ih} the mean of its row and the mean of its column, and add the mean of all elements of **E**.
4. Analyze **F** by principal components.

Rohlf (1968) found that this method represents distances between entities more accurately than principal-component analysis. A computer program may be obtained from Rohlf, Kishpangh, and Kirk (1971).

Multidimensional Scaling

According to Shepard (1980), the idea that stimuli can be mapped by points in space so that similarity is represented by spatial proximity goes back to Isaac Newton. Newton suggested that spectral hues could be represented on a circle; Henning thought odors could be represented within a prism. In 1952 Torgerson developed a workable method of metric multidimensional scaling, but the procedure involved long, complex operations and assumptions that are difficult to satisfy. It was not until Shepard worked out his procedure in 1961 that a general methodology, much different from factor analysis, became available.

Shepard (1962) called his scaling approach the analysis of *proximities.* He used an iterative approach to adjust the position of points in a space until the rank order of the interpoint distances was, as nearly as possible, the inverse of the rank order of the corre-

sponding similarities. He then sought the space of the smallest number of dimensions for which the departure from a perfect ranking was acceptably small. The qualitative, ordinal relations in the similarity data turned out to be sufficient to determine the quantitative metric structure of the spatial representatives. J. B. Kruskal (1964) subsequently developed a measure of departure from the monotonic relation posited between similarity and distance. This measure, which he called a stress measure, is given by

$$S^2 = \left[\frac{\Sigma(d_{ij} - \bar{d}_{ij})}{\Sigma d_{ij}^2}\right]^{\frac{1}{2}} \tag{4-4}$$

Here the d_{ij} are the distances between the points of any particular iteration given, in terms of the $N \times K$ coordinates X_{ik} of the N points in the K-dimensional euclidean space. The $\bar{d}_{ij}$ are numbers that are monotonic with the similarity data S_{ij} and minimize stress relative to the spatial distances d_{ij} at each interaction. To analyze the matrix of data for each subject separately, a metric method of individual difference scaling (INDSCAL) was developed by Carroll and Chang (1970).

In nonmetric multidimensional scaling the elements of the $N \times N$ matrix of observed similarities are ordered by rank from the smallest to the largest similarity coefficient. The indices may be distances or correlations, or they may express rank orders of the dissimilarities. Coordinates for the reduced space are computed by means of the widely employed MDSCAL program. Rohlf (1972) reports that multidimensional scaling seems better than principal-component analysis in accounting for both the larger intercluster distances and the fine differences between numbers of a cluster. A recent review by Green and Carmone (1970) is helpful. There are also two volumes on multidimensional scaling (Shepard, Romney, and Nerlove, 1972); the first volume covers theory and the second deals with applications.

One of the first applications of the MDSCAL program was to a set of judged similarities between 14 spectral colors. A two-dimensional circular fit to the 14 hues was obtained; reading clockwise it went from yellow to red to violet to blue to green. A fit by Carroll

and Chang (1970) using INDSCAL on other collected data revealed two orthogonal axes corresponding to a red-green dimension and a blue-yellow dimension.

Ordination in Entity Space

As indicated earlier, ordination in entity space begins with an $N \times N$ matrix **Q** of correlations. The **Q** matrix may be analyzed by principal-component analysis or principal factor analysis (communalities in diagonal cells of correlation matrix), but the method of analysis is the same. The matrix of factor loadings or the matrix of principal components can be rotated to simple structure. Instead of computing factor scores or component scores, however, the entity loadings on each factor are plotted. These are the coordinates of the entities in E-space.

In an illuminating discussion of the relation between test and person factors, Ross (1963) points out that setting entity means to zero and then factoring tests (*R*) shifts the origin to the centroid of the tests. Setting test means to zero and then factoring entities (*Q*) shifts the origin to the centroid of the entities. The data matrix yielding *Q* correlations is nearly always column-centered first and then row-centered. It follows from this that the *Q* factors (or components) are almost always bipolar. Nearly all will be characterized both by high positive and high negative loadings. Since the researcher defines each entity factor in terms of the entities with either high positive or high negative factor loadings, each factor defines two clusters rather than one. Thus if three factors are isolated, six clusters will be identified. Cognizant of this problem, Overall and Klett (1972) suggest the addition of a constant to shift the centroid. Guertin (1966) uses distances, which are unaffected by direction of difference in score profile, in his *Q* analysis. A third approach is to factor the matrix of profile cross-products after column-centering.

The problem with the last two methods is that the factors reflect not only profile shape but elevation and scatter as well. The investigator must therefore consider the consequences of factoring to locate entities in E-space. Using principal components in A-space appears to be the preferred procedure.

Ordination Studies

Ordination in attribute space is frequently used in biology and ecology but not often in the social sciences. Rohlf (1972) has reported on a comparison of nonmetric multidimensional scaling analysis, principal-component analysis, and Gower's principal-coordinate analysis. Nine different data sets were used: butterflies, frogs, pigeons, coral caiman, rats, bees, snails, mosquitoes, and nematodes. The groups consisted of 20 to 70 specimens with between 26 and 185 characteristics. The configurations of points obtained were quite similar for the three ordination techniques. Rohlf found that nonmetric scaling (MDS) gave better results as measured by the correlation between the distances in the k-dimensional configuration and the original distances. Principal-coordinate analysis gave a better fit than principal-component analysis when missing values were present. The MDS solution was judged to be superior unless there are a large number of entities.

An example of E-space ordination can be given for fighting ships. To find out how many types of fighting ships could be identified, Cattell and Coulter (1966) collected data from *Jane's Fighting Ships* (1964–1965). They used 12 measures to describe 33 vessels: displacement; length; beam; number of light, medium, heavy, and very heavy guns; number of personnel; maximum speed; submersibility; continuity of deck construction; and number of planes carried. These measures were then converted into standard-score form. The authors computed the distances between the 33 ships and then applied a multidimensional scaling analysis (INDSCAL). The two-dimensional plot in Figure 4-2 offers substantial evidence of the value of multidimensional scaling as an ordination procedure.

The analysis revealed that the ships could be conceived as embedded in two-dimensional space. The ship positions were defined by their coordinates on Axis I and Axis II. The ship code numbers are as follows: light cruisers, 1–5; heavy cruisers, 6–8; battleships, 9–13; aircraft carriers, 14–18; destroyers, 24–28; and frigates, 29–33. Visual inspection shows the heavy cruisers (upper right) cluster together, but of the aircraft carriers only 16 and 17 are close. The light cruisers group together where the two axes intersect. The destroyers and frigates separate out but are close neigh-

Figure 4-2. A Two-Dimensional Mapping of 33 Fighting Ships by INDSCAL.

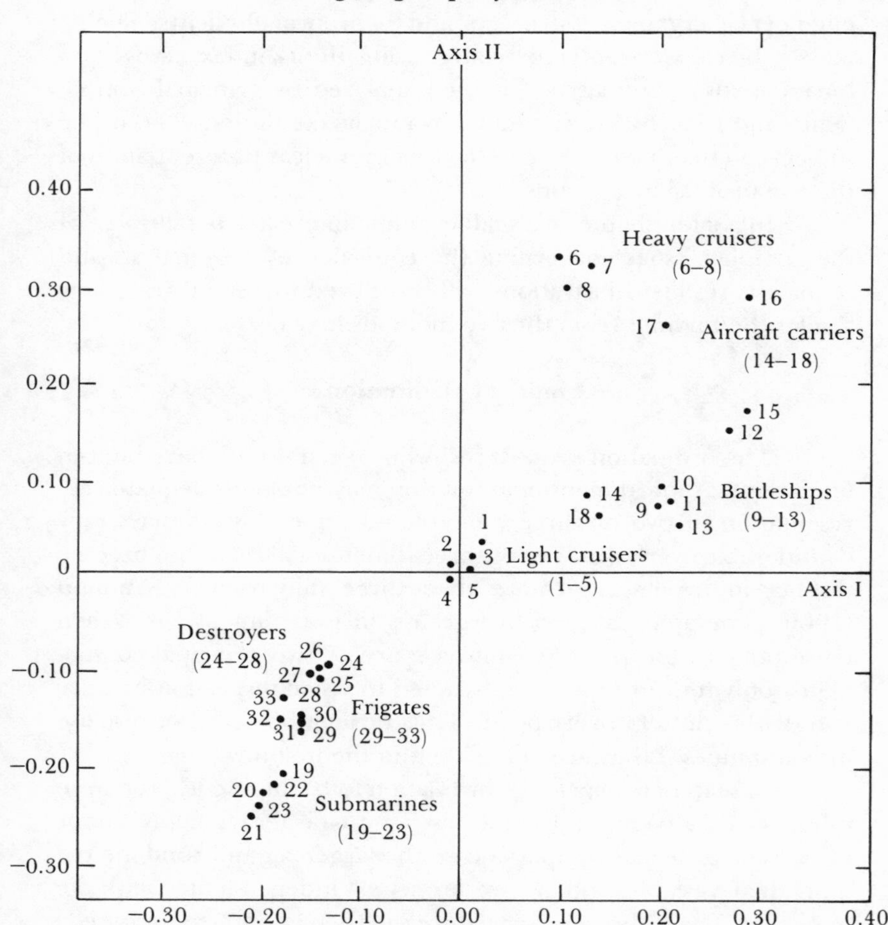

bors. The submarines appear as a discrete group. Four of the battleships (9, 10, 11, and 13) cluster together, but some aircraft carriers (14 and 18) are close to the battleships. Ships 12 and 15 seem unique. Thus the INDSCAL provides an effective visual separation of the various types of ships except for the aircraft carriers.

Another study, more complex, has been reported by McClung (1963). The aim was to isolate occupational personality

types on the basis of biographical data, attitudes, needs, and temperament. She selected 37 highly successful life insurance salesmen, engineers, clergymen, journalists, and theoretical physicists. The 37 male subjects were intercorrelated using the *G* index across 653 binary items. The matrix was then analyzed by principal components, and 13 factors were rotated to a varimax solution. Plots of the subjects on the first five E-space factors show clear differentiation of the five professional groups.

In a later chapter we shall examine in greater detail some of the complex issues concerning the equivalence of *Q*-analysis and *R*-analysis results. Illustrations will be offered to demonstrate how similar the findings from the two methods may be.

Limits of Ordination

The ordination procedures, while often useful, have built-in limitations. A low-dimensional solution may not give adequate representation in two or three dimensions. Some clusters may show considerable overlap in two- or three-dimensional space but become distinct in hyperspace (more than three dimensions). Sammon (1969) generated data for five groups in four dimensions. When these points were projected into a space of two principal components, only four groups were disclosed to visual inspection because two of the clusters overlapped. Thus ordination may not disclose discontinuities if they are not evident in the first few dimensions.

Researchers may judge ordination to be impractical for large numbers of attributes. In this case one can always apply factor analysis or principal components to the larger set and combine the individual variables into a smaller set of independent composite measures. The ordination procedure can then be applied to correlations or distances based on the composite measures. A related issue concerns the presence of a large general factor in the data. When a general factor is present, the coefficients of the first principal component (or first factor) are all large and positive, at least 20 percent of the variance is accounted for, and variable intercorrelations are substantially all positive. The problem is that such a factor tends to obscure profile patterns present in subsets of entities. Vocational-interest items, for example, often require responses of like-indiffer-

ent-dislike. Such items tend to elicit a response bias favoring a like response that is fairly uniform across items. Another example is the CL-90 (Derogatis, 1977), a complaint inventory that calls for ratings of complaints on a five-point scale of distress from not at all to extremely. Many patients exhibit an acquiescent response that results in a sizable general factor. One possible solution to this problem is to subtract an estimate of this confounding dimension from the obtained scores.

Summary

This chapter describes and explains some of the methods used to map entities into a space of two or three dimensions. The process, called ordination, can locate entities in attribute space or in entity space. Ordination in A-space is shown to involve a dimensional analysis of attribute intercorrelations, or *R* analysis. In E-space, ordination involves an analysis of entity intercorrelations called *Q* analysis. Three models of dimensional analysis—maximum variance, factor analysis, and multidimensional scaling—are described and contrasted. Principal components summarize the data in the form of new dependent variables. Factor analysis explains the observed correlations in terms of a smaller set of hypothetical underlying variables, both common and unique. A modified form of principal components called principal coordinate analysis is also presented. Multidimensional scaling is a method for representing a matrix of proximities by a configuration of points in low dimensional space. Each procedure is outlined, and several ordination studies, both in A-space and E-space, are summarized as illustrations of these approaches. The chapter concludes with a discussion of the limits of ordination such as too many dimensions, too many attribute variables, and the presence of a strong general factor.

The next chapter is devoted to single-level nonhierarchical clustering techniques. The principal methods described are single, complete, and average linkage as well as iterative partitioning methods including *k* means.

CHAPTER 5

Single-Level Cluster Methods

This chapter offers several definitions for the term cluster. Two kinds of clusters, the compact and the chained, are differentiated and related to various methods of cluster analysis. We begin by differentiating single-level nonhierarchical and multilevel hierarchical techniques. The single-level methods are subdivided into those that isolate single clusters successively and those that iteratively partition a sample into multiple clusters. After discussing the single-cluster search techniques (single linkage, complete linkage, and average linkage), we shall examine a variety of techniques that form multiple clusters by partitioning data sets. The density or mode-seeking methods are outlined here as well as several less popular techniques.

The Concept of Cluster

Although there is little general agreement on what constitutes a cluster, it is generally accepted that there are several cluster structures. It has been suggested (Rice and Lorr, 1969) that chained and compact clusters warrant separation. The *chained* cluster is defined as a category of entities in which every member is more like *one* other member than it is like any entity not in the category. The

compact cluster is defined as a subset of entities in which all members are more like *every* other member than they are like entities in any other subset. Cattell and Coulter (1966, p. 239) describe two kinds of clusters that appear very similar to the two just defined: a homostat, which is a set of entities "standing at closely similar positions in space;" and a segregate, a set of entities continuously related through other entities in the subset and isolated from those outside but not necessarily similar in position. Thus the definitions of chained cluster and the segregate match, as do the definitions of compact cluster and the homostat.

A spatial or geometric conception of clusters is helpful here. The entities can be regarded as points in n-dimensional space, where each of the n attributes is represented by a separate axis. The location of each entity is specified by its coordinates on the n axes. Clusters can then be conceived as regions of high density separated by regions of low density. This view of density and separation is evident in two well-known definitions. Jardine and Sibson (1971, p. 266) define a cluster as "a set of objects characterized by the properties of isolation and coherence." Gengerelli (1963, p. 458) defines a cluster as "an aggregate of points in test space such that the distance between any two points in the cluster is less than the distance between any point in the cluster and any point not in it." The definitions given by McQuitty for single-link clusters (1957, p. 209) and for complete-link clusters (1961, p. 677) correspond closely to those given for the chained and compact types.

The two kinds of clusters are illustrated in Figure 2-1 in Chapter Two. The compact cluster is roughly circular or spherical; the chained cluster is straggly, serpentine, or amoeboidal. The similarity relationships that define members of chained clusters are ordinal, asymmetric, and transitive. Members of compact clusters bear symmetric, transitive relations to each other.

Classification of Cluster Methods

Clustering procedures too can be classified to provide a conceptual scheme. A primary division is between methods that lead to mutually exclusive clusters and those that yield overlapping clusters. With but few exceptions cluster methods are designed to search for

mutually exclusive classes. The scheme that follows for mutually exclusive subsets resembles classifications proposed by Sneath and Sokol (1973), Anderberg (1973), Everitt (1974), and Bailey (1974).

Single-level nonhierarchical techniques are divisible into (1) the successive single-cluster search techniques that form clusters one at a time without iteration, (2) the iterative partitioning techniques that sort the data into multiple clusters by seeking to optimize some predefined criterion, and (3) the density or mode-seeking techniques that search for regions of high density and multiple modes in the data.

The *multilevel hierarchical techniques,* which yield treelike structures, include the *agglomerative* and the *divisive.* The agglomerative methods start with a disjoint set of entities and merge them by certain rules into fewer and more inclusive clusters until all are combined into a *conjoint set*—that is, a single, complete set. The divisive techniques begin with the conjoint set and partition the sample into smaller and smaller subsets.

Single-Linkage Analysis. Single-linkage analysis was first proposed by Florek (1951); it was also proposed independently by Sneath (1957) and by McQuitty (1957). A *link* is defined as the smallest distance between an entity and any other entity in the set. If correlations are used, a link is the largest correlation between an entity and any other in the set. To begin a cluster, search the matrix for the closest pair as a nucleus. Then add the entity that is closest to either member. Search the matrix again to find the entity that is closest to one member of the cluster and add that entity to the cluster. The cluster is thus extended to all possible entities that are continuously linked together via at least one member. Single linkage leads to long extended chains. The collection satisfies the definition of a chain cluster as a class of entities in which each member is more like some one other member than it is like any entity not in the class. Single-link chains are shown in Figure 5-1*a*. The circles represent the entities while the arrows indicate the relationship between them: Asymmetric ordinal relations are indicated by single-headed arrows; symmetric transitive relations are shown by two-headed arrows. The usual procedure is to begin with what McQuitty calls a reciprocal pair—that is, two entities that are mutually closest to each other.

Figure 5-1. Single-Link and Complete-Link Clusters.

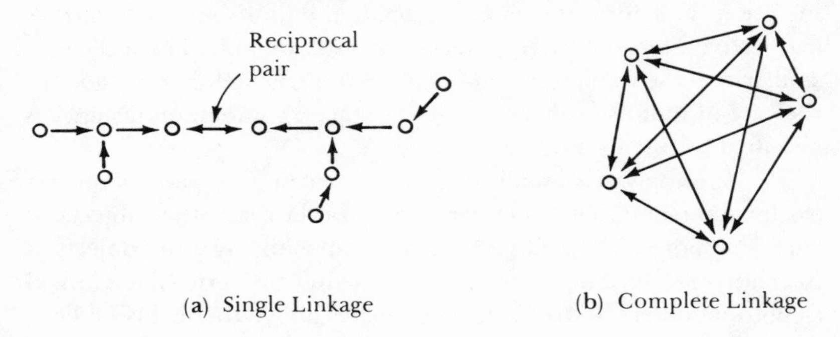

(a) Single Linkage (b) Complete Linkage

The cluster that emerges represents a chained or continuously connected subset. Entities at one end of the extended cluster resemble each other more closely than entities at the other end of the cluster.

In McQuitty's original report (1957) the method was illustrated with an analysis of the correlations among responses of 15 persons who rated themselves on 121 introversion-extroversion items assembled by Stephenson (1953). Application of the procedure yielded two distinct clusters of introverts and extroverts. Because the underlying clusters were compact the method successfully separated the two subgroups present. But most social scientists want compact clusters and thus tend to use complete or average linkage. The strongest advocates of the single-link method are biologists who apply it in hierarchical clustering, which is discussed in the next chapter.

Complete-Linkage Analysis. The method of complete-linkage analysis was proposed independently by workers in several fields. Horn (1943) applied the method to cluster test variables. Sorensen (1948) developed the method for use in ecological studies. McQuitty suggested typal analysis (1961) and rank order typal analysis (1963). The type generated is defined as a group of entities in which each member is more like every other member than it is like members of

any other type. To generate a cluster it is necessary to specify some inclusion limit for admission. Typically the limits set are arbitrary (subjective), but objective criteria are also available. For indices of similarity some minimum value, say $r = 0.65$, is set for admission. In the case of indices of dissimilarity (distance), a minimum distance is specified for inclusion.

The analysis is usually begun by searching for pairs of objects (reciprocal pairs) that are closer to each other than other objects or correlate more highly with each other than with any other objects in the matrix. A third object is added to the nucleus of two if it is linked to both members—that is, if it correlates above the inclusion limit with both members. The process continues until no other objects can be added. Members of a complete-linkage cluster bear symmetric relations to each other; all members are mutually alike. Figure 5-1*b* presents a two-dimensional illustration of a complete-linkage cluster.

McQuitty (1963) has proposed a rank order typal analysis in which the threshold for inclusion is objective. The method requires every member of the cluster to be more highly correlated with every other member than with any member not in the cluster. All similarity coefficients in each column are ranked. A cluster must not include a rank higher than the number of entities in the cluster. If the rank of a similarity index between two objects exceeds the number of objects in the cluster, the requirement is not met. Saunders and Schucman (1962) have developed an objective complete-link procedure called syndrome analysis that operates on squared distances between entities. Saunders first identifies all mutually closest pairs and then all mutually closest triplets, quadruplets, and higher orders. In the last step subsets within larger subsets are eliminated.

Considered spatially the complete-linkage procedure generates sectors of equal area around the point of origin in the attribute space. Each sector, like a slice of pie, subtends an angle within the circle whose cosine (which equals the correlation) conforms to the inclusion limit specified. The clustering procedure thus segregates or groups together all points within the sector. Since the sectors may overlap, the clusters will overlap if the inclusion limit is too broad. With distance measures, circles of radius $D/2$ are generated around entity pairs in centers of high density in the data. These conditions

point toward a defect inherent in single-level complete-linkage analysis: overlapping clusters. If the aggregation of points is fairly evenly spaced, the subsets represent arbitrary partitions of points in space. Syndrome analysis avoids the problem of overlap.

Average-Linkage Analysis. Because of the stringent requirements of complete linkage, the clusters tend to be small. A less stringent and perhaps more realistic method is to average the indices of similarity on the basis of which an entity is admitted to a cluster. A cluster is then defined as a subset of entities in which each member is, on the average, more like every other entity than it is on the average like any entity in any other cluster. Clustering by average linkage seems to have been first proposed by Sokal and Michener (1958) and popularized by Sokal and Sneath (1963). It should be noted, however, that the method was developed primarily for hierarchical agglomerative cluster analysis and not for single-level clustering.

The algorithm begins by listing for each entity all entities that correlate with it at or above a specified inclusion limit. Since the entity with the longest list is likely to be near the center of a cluster, it is selected as a pivot and combined with the entity most highly correlated with it. To this nucleus is added the entity that correlated highest on the average with the pair. The entire matrix is then searched repeatedly and the entity profile added that correlates highest on the average with those already in the cluster and yet exceeds the threshold. The procedure continues until there are no profiles whose mean correlation with cluster members equals or exceeds the inclusion limit.

Average linkage, like complete linkage, is likely to result in overlapping clusters. To eliminate overlap by excluding outliers and intermediate points (representing entities) between regions of high density, an exclusion limit can be specified. The cluster process, called BUILDUP (Lorr, 1966; Lorr and Radhakrishnan, 1967), uses correlations, but distances can be converted into correlations via the formula previously noted ($D^2_{ab} = 2 - 2Q_{ab}$). Membership in a cluster is determined by inclusion and exclusion cutoffs that are specified in advance. Beginning with a nucleus of three entity profiles, the matrix is searched for the profile that on the average correlates highest with the nucleus and satisfies the inclusion cutoff C_{in} (say $r = 0.60$). The process continues until no more profiles can be added to

the cluster. Next an exclusion cutoff C_{ex} is set that defines dissimilarity, excludes outliers, and prevents cluster overlap. Any profile in the residual matrix that correlates on the average C_{ex} or higher (say $r = 0.50$) with the first cluster is deleted. Profiles excluded are not considered again for inclusion in subsequent cycles. The second cluster is generated from the reduced matrix in the same manner. Profiles that satisfy the inclusion limit are added to a nucleus. Again the exclusion limit C_{ex} is applied to delete profiles that are too close to the second cluster. The process is continued until no cluster with at least four members can be formed.

The recommended inclusion limit is the value at which a coefficient of correlation based on k independent variates is significant at $p = 0.025$. The exclusion limit is set at $p = 0.050$. Experimental trials with simulated data have shown these values to be optimal (Suziedelis, Lorr, and Tonesk, 1976). The assumption is made that the measures in a profile represent a random sample from a specified universe. Should the measures be factor scores that are relatively uncorrelated, these rough criteria work fairly well.

Blashfield (1976b) reports that the method has been applied widely; applications may be found in Goldstein and Linden (1969), Nerviano (1976), and Lorr (1966). Empirical comparisons indicate that BUILDUP and the minimum-variance technique (Ward, 1963) yield quite similar results, especially if correlations are converted into distances for Ward's procedure.

The effectiveness of average linkage with an exclusion limit is demonstrated in a study of MMPI (Hathaway and McKinley, 1951) by Lorr and Suziedelis (1982). Gilberstadt and Duker (1965) and Marks, Seeman, and Haller (1974) derived a series of profile types on the basis of their frequency of occurrence in successive samples of psychiatric patients. In their coding system, the profiles are labeled in order of their scale elevation from highest to lowest beyond the limits of a T score of 70. The 35 profiles, combined from these two sources, are expressed in terms of 13 scale scores. In the initial step, each of the 13 scale-score distributions was transformed into standard-score form with a mean of zero and a standard deviation of 1. The clustering algorithm was applied with correlations of 0.51 and 0.44 as inclusion and exclusion cutoffs.

The cluster analysis disclosed four profile types or clusters.

Figure 5-2. Mean MMPI Profiles of Four Clusters of Gilberstadt/Duker and Marks/Seeman/Haller Code Types.

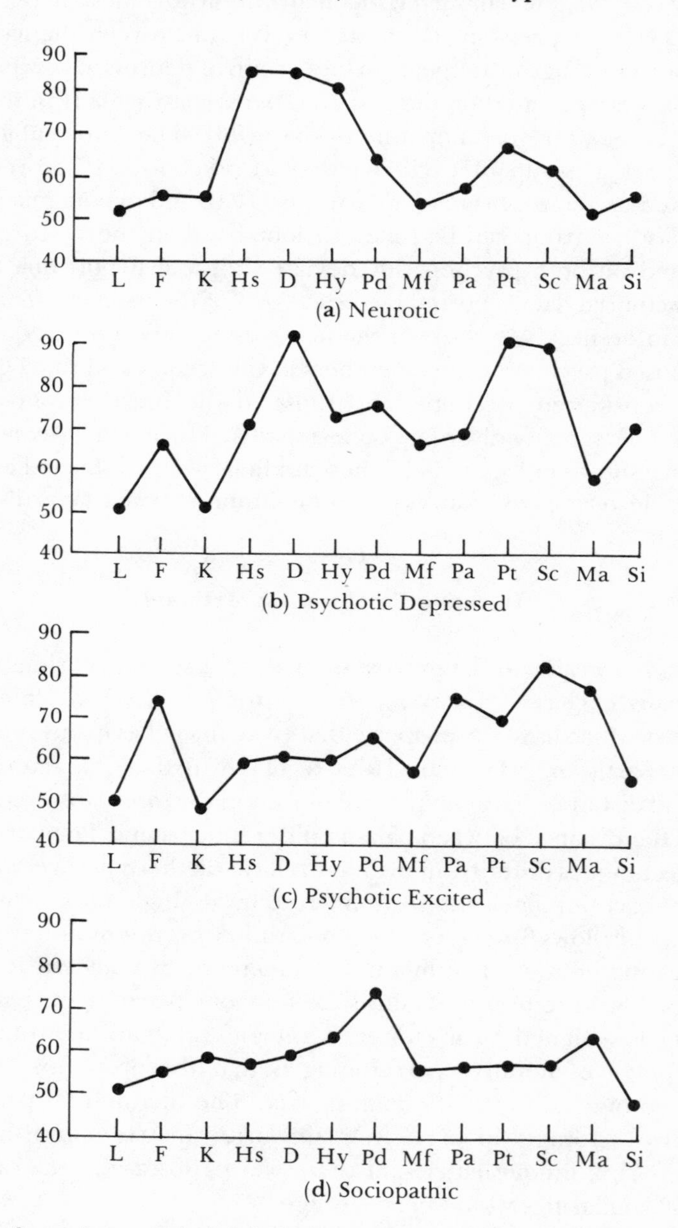

Source: Lorr and Suziedelis, 1982. Used by permission.

The first subgroup is called neurotic because the mean T scores above 70 include characteristic neurotic syndromes: hypochondriasis (Hs), depression (D), hysteria (Hy), and psychasthenia (Pt). The second subgroup appears to be a group of depressed psychotics; T scores are elevated on depression (D), psychasthenia (Pt), schizophrenia (Sc), and psychopathic deviate (Pd). The third subgroup comprises a group of excited paranoid psychotics; T scores are elevated on schizophrenia (Sc), paranoia (Pa), and hypomania (Ma). The last subgroup can be called sociopaths; their mean scores are elevated on both psychopathic deviate (Pd) and hypomania (Ma). The actuarial data reported by the authors of the code type confirm these inferences. Members of the first cluster were most frequently diagnosed psychoneurotic, members of the second and third group were considered psychotic, and those in the fourth group were labeled personality disorders or sociopaths. The mean T-score profiles are shown in Figure 5-2. These findings were confirmed by two other clustering procedures called minimum variance (Ward's) and *k*-means.

Iterative Partitioning Methods

Partitioning is the process of dividing a set of N entities into g mutually exclusive clusters. The resulting *partition* is a "family of clusters which have the property that each object lies in one member of the partition" (Hartigan, 1975, p. 11). A further characteristic is that all distances between pairs of objects in the same cluster are less than the distance between pairs in different clusters. The partitioning techniques differ from the hierarchical methods in several ways. First, partitioning leads to nonhierarchical single-rank solutions; second, it allows for correction of poor initial partitions by iteratively relocating entities. The hierarchical method, by contrast, leads to multilevel structures and allows for only one assignment. Once an entity is assigned to a cluster it remains in that subgroup. An advantage of iterative partitioning is that the programs require storage only of the $N \times n$ data matrix. The hierarchical program requires storage of a $N \times N$ similarity matrix consisting of $N(N-1)/2$ unique entries. If there were 100 cases, for example, 4,950 similarities would need storage.

In optimal partitioning a set of cases is iteratively partitioned to maximize some predefined criterion function. The goal is to find the partition that is in some sense best. The most direct way to find the best solution would be to form all possible partitions of N entities and then choose the best one. The difficulty is that the number of partitions even for moderate N is so enormous that the task even for a fast computer is not feasible—for example, there are 2,042 ways of assigning 12 entities to two groups, 86,526 ways to three groups, 611,501 ways to four groups, and 1,379,400 ways to five groups (Lyerly, 1968). Since all possible partitions cannot be exhaustively analyzed, procedures have been developed to sample a small subset of the possible solutions.

Criteria for Optimal Clusters

Many clustering criteria are designed to minimize the scatter or variation within clusters and to maximize the variation between clusters. Nearly all such criteria derive from the basic matrix equation

$$\mathbf{T} = \mathbf{W} + \mathbf{B} \tag{5-1}$$

where **T** is the total dispersion matrix (total sum of squares and cross-products), **W** is the pooled within-group dispersion matrix, and **B** is the between-groups dispersion matrix (sum of squares and cross-products). The matrix **W** can be further defined by $\mathbf{W} = \Sigma^g W_i$. Since **T** is fixed, to minimize **W** is equivalent to maximizing **B**. For $n = 1$, Equation (5-1) reduces to a scalar equation basic in analysis of variance. The ratio B/W is known as the **F** ratio when multiplied by the degrees of freedom. The following list presents some of the principal criteria that have been developed. Reviews may be found in Anderberg (1973), Everitt (1974), and Friedman and Rubin (1967).

1. *Minimize the trace of* **W**. One criterion frequently proposed is to minimize the trace of **W** over all partitions into g groups. The trace of a square matrix is simply the sum of its diagonal elements. In **W** these values constitute the pooled within-group

sums of squares. By this procedure all the entities in a cluster are closer to their own cluster centroid (mean) than to the centroid of any other cluster. Since trace **T** = trace **W** + trace **B**, to minimize trace **W** is equivalent to maximizing trace **B**, the between-group sums of squares. Such a criterion has been proposed by Forgy (1965), Jancey (1966), MacQueen (1967), and Ball and Hall (1965).

2. *Maximize the ratio* $|\mathbf{T}|/|\mathbf{W}|$. This ratio of two determinants can be expressed as

$$\frac{|\mathbf{T}|}{|\mathbf{W}|} = |\mathbf{I} + \mathbf{W}^{-1}\mathbf{B}| \tag{5-2}$$

The left-hand side of the equation is a ratio of two scalars. Its reciprocal, $|\mathbf{W}|/|\mathbf{T}|$, represents the well-known Wilks lambda (Λ) for testing whether two or more groups differ in location (mean values). But the criterion to be maximized is $|\mathbf{T}|/|\mathbf{W}|$. When the ratio is a maximum, the associated partition is selected as best. This criterion, proposed by Friedman and Rubin (1967), is discussed later in this chapter.

3. *Maximize the trace of* $\mathbf{W}^{-1}\mathbf{B}$. Another criterion, suggested by Friedman and Rubin, is to maximize the trace of $\mathbf{W}^{-1}\mathbf{B}$ over all partitions into g groups. Here $\mathbf{W}^{-1}$ represents the inverse of matrix **W**. Its value can be shown to maximize the sum of eigenvalues

$$\text{Trace } \mathbf{W}^{-1}\mathbf{B} = \Sigma\lambda_g \tag{5-3}$$

The eigenvalues λ are solutions to the determinantal equation $|\mathbf{B} - \lambda\mathbf{W}| = 0$.

4. *Maximize the largest eigenvalue of* $\mathbf{W}^{-1}\mathbf{B}$. This criterion is known as Roy's largest-root criterion.

As Anderberg (1973) points out, there are problems associated with criteria 2, 3, and 4. First, by involving computation of eigenvalues at each stage they limit the number of variables that can be used; moreover, the need to compute the inverse of a matrix that may be singular adds to the cost of computation. Second, Friedman

and Rubin observe correctly that there are no guidelines other than empirical studies for making a choice between the criteria.

In his review of computer programs that perform cluster analyses, Blashfield (1976b) describes eight that perform iterative partitioning analyses. Seven of these programs are described in Appendix A. A number of characteristics distinguish these programs: the statistical criteria for optimality, the method of initiating partitions, the type of move or pass used to assign entities to clusters, determination of the number of clusters, and the allowance made for outliers. Since the statistical criteria for optimal clusters have been discussed, we turn now to the processes involved in partitioning a data set.

Three distinct processes are likely to be involved in a partition: a method of initiating clusters, a method of allocating new entities to the initial clusters, and a method of determining when further reallocation of entities to the established clusters should be stopped. Several closely related methods of partitioning were proposed independently and almost simultaneously by Forgy, Jancey, MacQueen, and Hall and Ball. All require specification of the number of groups (k), all require definition of the group mean, and all use euclidean distance as the measure of similarity. A common tactic is to make an initial partition of the entities and then alter cluster membership iteratively. Another approach is to use estimates of cluster centroids (called *seed points*) around which clusters may be formed.

Method of Initiating Clusters. A set of k means or centroids can be used as cluster nuclei or seed points. Some of the methods are the following:

- Use the first k units in the data set (MacQueen, 1967). While this procedure is simple, Monte Carlo studies suggest that it has not proved to be effective (Milligan, 1980).
- Number the entities, assign random numbers, and select a random set of k entities.
- Use *a priori* hypotheses to specify k entities from the data set.
- Partition the entities into k mutually exclusive groups and compute the group centroids for use as seed points (Forgy, 1965).
- Choose k seed points that appear to span the data set.

Type of Move or Pass. The problem is how to assign entities to particular clusters in order to minimize within-cluster variance. The best-known method is the *k-means pass:* Assign each entity to the cluster (of *k* clusters) with the nearest mean. The second type of move, the *hill-climbing pass* (Friedman and Rubin, 1967), will be described shortly. A procedure due to Forgy (1965) is called a *reassignment pass.* Starting with the best partition currently known, each entity is reassigned to the cluster with the nearest centroid. The values of the newly formed partition are then calculated. Friedman and Rubin also have a *forcing pass,* but in their trials this technique was only occasionally helpful. The procedure starts with the best partition known and assigns each entity to an outside cluster with the nearest center or centroid. After all the entities in a group have been dealt with, the process is applied to the next cluster.

Fixed vs. Variable Number of Clusters. The manner in which the program determines the final number of clusters varies with the method. In Anderberg, CLUS (Friedman and Rubin), and MIKCA (McRae) use is made of a prespecified number whereas BCTRY and ISODATA allow for a variable number. ISODATA is most flexible in this regard.

Problem of Outliers. The algorithms in CLUSTAN, CLUS, ISODATA, and BCTRY assign entities outside a specified distance from any cluster centroid to a group of *outliers.* These can be added to a cluster on a later pass if the entity is within the inclusion limit. An outlier is any case in a relationship that is not part of the majority of cases—that is, the case lies well above, below, or outside the relationship. In Chapter Eight it will be seen that outliers can seriously distort cluster structures.

Forgy's k Means. Forgy (1965) suggested a simple algorithm that falls under the *k*-means concept and illustrates the partitioning process. The steps are: (1) Begin with an initial partition of the data set into a number of clusters; compute the mean of each cluster. (2) Allocate each entity to the cluster with the nearest mean. (3) When all entities have been assigned, compute the new centroids of the clusters. (4) Alternate steps 2 and 3 until no data points change membership.

MacQueen's k Means. The procedure developed by MacQueen (1967) partitions a sample of *N* entities into *k* sets on the basis of a

euclidean distance measure. The partitions are reasonably efficient in the sense of within-class variance. The program starts with a user-specified value of k (number of groups). To sort the N data units into k clusters, three basic steps are followed: (1) Take the first k entities as the first k clusters. (2) Assign the remaining $N - k$ entities (one after the other) to one of the k clusters on the basis of the shortest distance between the entity and the centroid (mean) of the cluster. After each assignment the centroid is recomputed (updated) for the cluster gaining a member. (3) After each entity has been assigned to one of the k clusters, take the cluster centroids as fixed seed points. Each data unit is again assigned to one of the clusters according to the closest distance to a centroid. The process can be repeated, and the series of partitions thus produced will converge with decreasing within-class variance. For a useful explanation of convergence see Anderberg (1973, pp 165–166).

The ISODATA k Means. Perhaps the most elaborate of the nearest-centroid clustering methods is called ISODATA. It was developed by Ball and Hall (1965) at the Stanford Research Institute. A version of the method has been incorporated into what is called the PROMENADE system, an on-line data analysis package using interactive graphics. Anderberg (1973) says that the method is clearly the most expensive of the nearest-centroid methods. The method involves six basic steps: (1) Choose a set of seed points (or generate one). (2) Assign each entity to the cluster with the nearest seed point. These remain fixed until the cycle is completed. (3) Recompute the cluster centroids and repeat the assignments to the nearest centroids until convergence is achieved. (4) Discard clusters fewer than a specified number of data units. (5) Perform a *lumping* (merger) of clusters if the number of clusters is twice the cutoff. (6) Perform a *splitting* of clusters if the number of clusters is half the cutoff or less. The question of when to stop iterating on the partition is a problem. With PROMENADE the subjective approach is part of the system. Mezzich (1982) found that ISODATA does not do as well as simpler k-means procedures, although it ranks sixth across three evaluative criteria when 18 procedures are compared (eight algorithms times three indices of similarity).

Milligan (1980) and others have shown that random seed points lead to poor recovery rates. This discovery led Milligan and

Figure 5-3. Four MMPI Profile Clusters Based on *k*-Means Program.

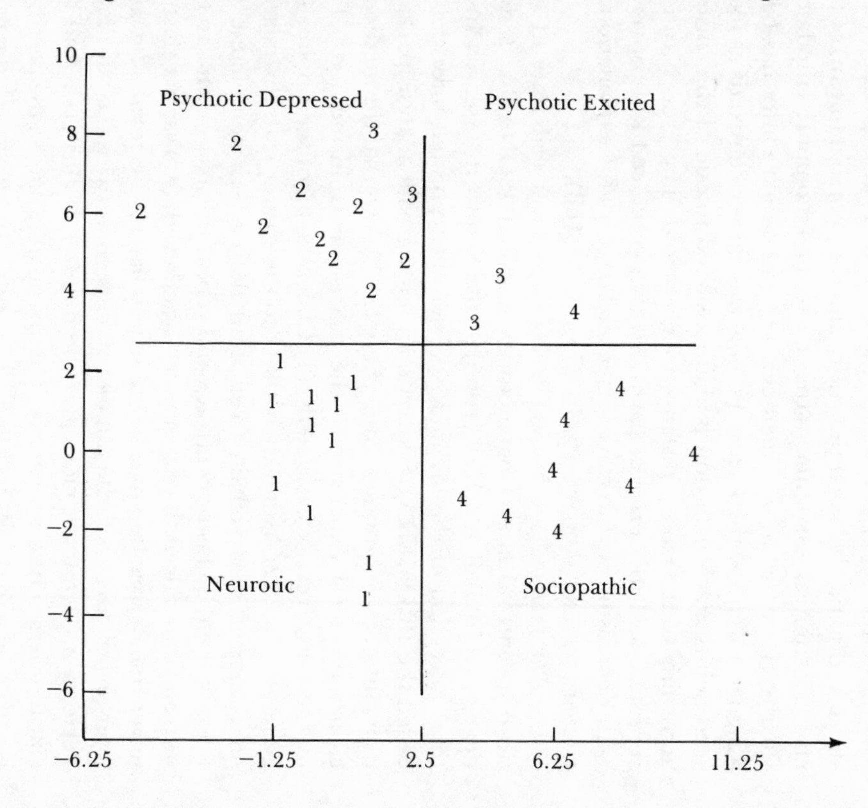

Sokal (1980) to develop a two-stage algorithm that possesses robust characteristics under several error conditions. The procedure begins with the group-average-linkage method of hierarchical analysis. The average-linkage procedure was selected because it was found to produce the best recovery of cluster structure under perturbing error conditions. The next step is to determine the number of clusters to assume. Then the cluster centroids are used as seed points in a variant of the *k*-means algorithm. To estimate the number of clusters, Milligan generates random noise data sets via a multivariate random number generator. The data sets are then analyzed by the clustering algorithm and a biserial test statistic is derived for each partition to determine the number of clusters to retain. A limitation of the algorithm is the need to test for the number of clusters via random data sets. An alternative procedure is to apply Mojena's Rule One to determine the number of clusters; this rule is described in the next chapter.

An Illustration of k Means. To illustrate the *k*-means procedures, the Gilberstadt/Duker and Marks/Seeman/Haller MMPI code types described earlier in the chapter were used as a data set (Lorr and Suziedelis, 1982). The standardized scale scores were first clustered by Ward's hierarchical procedure, which is outlined in the next chapter. Mojena's Rule One (1977) indicated the presence of four clusters. The centroids of the cluster member profiles were used as seed points in the BMDP-79 *k*-means program. The location of members of the four clusters established is shown in Figure 5-3. The points were generated by computer onto a plane through the center of the clusters. The BUILDUP average-linkage procedure assigned 24/35 of the code types to the four clusters; *k* means allocated all 35 to the four. The correspondence between average linkage and *k* means were close.

Friedman/Rubin Methods

A set of invariant criteria for grouping data has been developed by Friedman and Rubin (1967). Their approach was to use computer procedures to obtain the best partition of N objects into g groups. Three of the criteria previously described were elaborated: (1) Minimize the trace of $\mathbf{W}$; (2) maximize the ratio of determinants

$|\mathbf{T}|/|\mathbf{W}|$; and (3) maximize the trace of $\mathbf{W}^{-1}\mathbf{B}$. The principal technique used in conjunction with the criteria was the hill-climbing pass. The process can be described in the following way. Start with a given partition into g groups. Consider moving an entity to another group. If no move will improve the criterion, the entity is left where it is; otherwise it is moved to increase the criterion. The second entity is processed in the same way, as well as the third, the fourth, and all the remaining entities. This procedure is applied once to each entity in a specified order. After several hill-climbing passes, the point is reached at which no move of a single entity will increase the criterion. At this point a local maximum is reached. Friedman and Rubin also describe *forcing passes* and *reassignment passes.*

Friedman and Rubin tested their procedure on Fisher's iris data (1936), on artificial data, and on a sample of bee data. Fisher's iris data consisted of three species (*Iris setosa, Iris versicolor,* and *Iris virginia*); there were 50 plants of each species and four measurements on each plant. The computer-generated artificial data consisted of five groups of 10 objects; each object was characterized by a vector of five measures. The bee data consisted of 97 American species of four genera; each species was described on 97 measurements. A principal-component analysis yielded three components that were used to describe the species.

In their discussion Friedman and Rubin say that they prefer $|\mathbf{T}|/|\mathbf{W}|$ among the three criteria. This criterion is invariant under nonsingular linear transformations. Moreover, it has demonstrated a greater sensitivity to the local structure of the criteria than the other criteria. This result is fortunate since it is easier to compute $|\mathbf{W}|$ than $\mathbf{W}^{-1}$, which is computed only for the final output. The major fault of the minimum trace $\mathbf{W}$ criterion is that it does not take into account the within-group covariance of the measurements.

A number of applications of Friedman and Rubin's procedure have been reported. Paykel (1971) used it to classify depressed patients into four groups. Bartko, Strauss, and Carpenter (1971) found it costly in computer time and least effective in recovering a set of archetypes of psychiatric diagnostic categories. Everitt, Gourlay, and Kendell (1971) applied the method to validate traditional psychiatric syndromes. Mezzich (1982) conducted an empirical comparison of seven algorithms, three indices of similarity, and four data

sets. The data bases consisted of treatment environments, archetypal psychiatric patient groups, iris plants, and ethnic populations. Friedman and Rubin's procedure ranked first in differentiating the iris plants but not very well on the overall comparisons across the various criteria. If their method is used, the safest approach is to arrive at a starting partition by a simple procedure before applying the algorithm.

Multivariate Mixture Analysis

Most cluster analysis procedures measure the similarity between two entities and then group them in a way that maximizes within-cluster similarity. Wolfe (1970) has derived a system of cluster analysis without assumptions about similarity. Since entities within a type differ from one another, it is reasonable to assume the existence of a probability distribution of attributes for a population belonging to this type. Members of a different type will have a different distribution of characteristics. The combined population taken from all types will have a probability distribution that is a *mixture* of distributions. The problem is to determine the number of distributions, define the parameters of each, and determine the class to which each entity belongs. The component distributions are assumed to be standard statistical distributions with unknown parameters.

Wolfe has developed maximum-likelihood estimation procedures for mixtures of multivariate normal distributions. Estimates are made of the proportion of a mixture from a given type as well as their means and covariances. The likelihood equations are solved by an iterative process using *a priori* hypotheses of the number of types present. The closest analogy to a similarity is the probability of membership of a point (entity) in a cluster. Wolfe regards his programs NORMAP and NORMIX to be continuous versions of the discrete partitioning procedure of Friedman and Rubin (1967). As a test, Wolfe applied the procedure to the iris data described in the previous section. The data were correctly resolved into three classes with means close to those reported by Fisher. Wolfe noted the same classification errors that occurred when Friedman and Rubin's method was applied.

NORMIX differs from other cluster methods in that its classifications of entities are not mutually exclusive. Instead every entity has a probability of membership in several clusters. The characteristics of a cluster are determined by averaging the measurements of all sample entities in such a way as to weight the contribution of each entity by its probability of membership in the cluster. It should be noted that NORMAP is not a complete procedure because the maximum-likelihood equations have to be solved iteratively starting with a set of initial estimates of each cluster—and these initial estimates have to come from some other technique. Different initial estimates can give different results, but if there are two different solutions, one can decide which is best by computing their likelihoods for the sample and choosing the one that is greatest. To obtain initial estimates of the number of groups and their characteristics, Wolfe suggests using Ward's hierarchical procedure, which is included as a subroutine. In another study Wolfe (1971) applied NORMIX to Strong Vocational Interest Blank data on 113 occupational groups. Thirteen types were found and compared to two solutions obtained from Ward's hierarchical procedure. In Wolfe's judgment the NORMIX solution was superior to the Ward.

NORMAP assumes equal within-group variance-covariance matrices whereas NORMIX allows them to differ. One way to use these programs is first to apply NORMAP with initial estimates of the parameters and then apply NORMIX. Everitt (1974) reports that both procedures are badly affected by outliers. In the Mezzich study (1982), previously described, the NORMAP/NORMIX procedure ranked lowest among all procedures. Multivariate mixture analysis, like *k* means and Friedman and Rubin's procedure, appears useful primarily as a way to arrive at a final clustering solution after another algorithm has been applied to a body of data.

Density-Seeking Techniques

When entities are regarded as points in *k*-attribute space, it is natural to conceive of clusters as regions of high density separated by regions of low density. This concept, known as the *density search technique,* has been used by a number of methodologists. A procedure called the Cartet count has been suggested by Cattell and

Coulter (1966). Given a coordinate system with each entity represented by a point, convenient intervals are taken to define *cartets,* which in a two-dimensional map would be squares. In *k*-space, the cartets (named after Descartes) are hypercubes. The computer program can then count the number of entities in each cube. A significance level could be set for the cube relative to the average total density to make it possible to locate clusters. Tryon and Bailey (1970) include such a program in their *Cluster Analysis* as core O clusters. Each scale is arbitrarily sectioned into three broad categories of high, middle, and low. The entities are then sorted into sectors. When a minimum of, say, five entities in a core type is set, a number of core types emerge. The method is illustrated with data from the MMPI and Holzinger measures of intellectual aptitude. There are, of course, distinct limitations to the procedure. There is seldom an *a priori* basis for accepting the arbitrary sectors. In fact, some natural clusters may have been sliced through. Nor is there an objective criterion for deciding on the number of clusters present.

Another density search technique, called the taxometric map, has been proposed by Carmichael and Sneath (1969). Clusters are formed by single-linkage analysis, but special criteria are used to determine when additions to clusters should be stopped. One criterion is to terminate if the point is much further away than the previous point; this indicates a discontinuity. The procedure is to compute the drop in average point-to-cluster distance of the neighbor nearest any point in the cluster from the average of the previous successive single-link distance. Another criterion is the ratio of the minimum similarity between any pair of points already in the cluster to the minimum similarity of the prospective candidate and any point in the cluster. The procedure has an important limitation to those lacking in expertise: There are numerous parameters controlling the technique that must be chosen by the investigator.

Wishart (1969b) developed a procedure called mode analysis that derives from single-linkage analysis. It is designed to search for natural groupings of the data by making estimates of the density surfaces in the sample distribution. Use is made of a linkage parameter *k* and a density threshold *R* for defining the radius of a sphere. Using a sphere of radius *R* around each point, the computer counts the number of points within the sphere. Spheres containing *K* or

more points are called *dense;* others are termed nondense. Dense points are then fused or merged by single-linkage rules. By increasing the radius, nested sets may be generated. A program for mode analysis is available in CLUSTAN.

Graph-Theoretic Techniques

When entities are represented as points in k-dimensional space, some clusters are globular while others are strung out into serpentine shapes. Some clusters may be spaced out along a straight line or even take the shape of a circle or donut. As will be seen later, some hierarchical methods, called minimum-variance techniques, may seriously distort or segment these groupings. Graph-theoretic methods, however, provide one useful means for uncovering such data structures. A *graph* (Zahn, 1971) is composed of a set of points called *nodes* that are joined together by *edges* (lines) with some or all of the nodes. A *path* in a graph is a sequence of edges (lines) joining two nodes. A graph in which there is a path from any node to every other node is *connected.* A closed path is called a *circuit.* If there is an edge (line) connecting each pair of nodes directly, the graph is *fully connected.* A *spanning tree* of a connected graph of N nodes is any set of $N - 1$ edges that represents one and only one path between any pair of nodes; it links together all the nodes with a minimum number of edges. A *minimum spanning tree* (MST) is the shortest spanning tree that can be constructed from the given set of edges. Examples of graphs are presented in Figure 5-4.

J. B. Kruskal (1956) was perhaps the first to develop an algorithm for finding the minimum spanning tree. The method corresponds to single-linkage cluster analysis and consists in adding the shortest link between two entities, then the second shortest, and so on, but excludes circuits or cycles. Gower and Ross (1969) obtained single-linkage clusters from minimum spanning trees by successively eliminating links, the largest first. Zahn (1971) reviews graph-theoretic methods and offers a program called MSTCLUS.

In the social and behavioral sciences the search has mainly been for *compact* clusters. The ordinal relationship found in sociometric choices, preferential orders, and communication networks

Figure 5-4. Minimally Connected Graph (a) and Maximally Connected Graph (b).

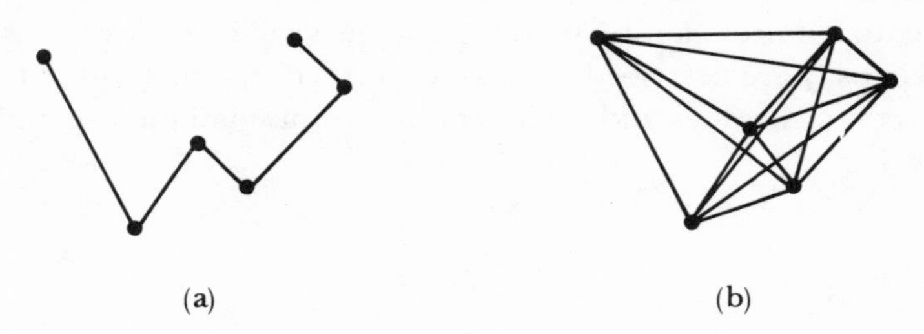

can be studied by means of single-linkage analysis, however, and the clusters found are those we have called chained or connected. Rice and Lorr (1969) found that the single-link algorithm is especially suited for isolating unfilled geometric figures like circles as well as rays emanating from a common origin.

Summary

In this chapter two kinds of clusters, the compact and the chained, are defined and differentiated. The compact cluster is roughly spherical while the chained cluster is serpentine or amoeboidal. The cluster methods are classified into the single-level nonhierarchical and the multilevel hierarchical. The single-level are subdivided into those that isolate single clusters successively and those that iteratively partition a set into multiple clusters. The successive single-level methods, which include single linkage, complete linkage, and average linkage, are outlined. The process of average-linking analysis is sketched and illustrated in some detail. Four statistical criteria for iterative partitioning of optimal clusters are examined and compared. Several methods for initiating clusters and several types of moves to optimize the allocation of an entity to a cluster are listed and compared. Three *k*-means procedures are sketched (Forgy, MacQueen, Ball & Hall). The Friedman-Rubin set of invariant criteria for grouping data are examined and evaluated.

Wolfe's multivariate mixture analysis methods which estimate the proportion of a mixture from a given class and its characteristics is presented. Three density search techniques which seek regions of high density are described. A final section is concerned with graph-theoretic techniques and their relation to the minimum spanning trees.

CHAPTER 6

Hierarchical Clustering Techniques

In the previous chapter we reviewed the nonhierarchical clustering methods for single-level structures. The present chapter is devoted to the hierarchical techniques, both agglomerative and divisive. After discussing the more popular agglomerative methods (single linkage, complete linkage, and average linkage), we shall examine Ward's minimum-variance technique as well as the centroid methods and, finally, the less frequently used divisive techniques for monothetic and polythetic clusters. We shall also consider a new divisive procedure called blockmodeling. The chapter closes with a splitting technique named AID for identifying nonoverlapping subgroups for predicting dependent variables.

Hierarchical Clustering

A hierarchy may be viewed as a family of nested multilevel classes. Hierarchical clustering techniques can be classified as *agglomerative* or *divisive.* The agglomerative methods begin with N entities that are sequentially merged at successive levels until all are included. The divisive procedures operate in the opposite direction — the entire collection is partitioned into finer and finer subsets at each level. The agglomerative methods build a *tree* or *dendrogram*

from branches to the root; the divisive methods begin at the root and form a branching sequence. The two-dimensional tree diagram shows the fusions or the partitions. Hartigan (1967) has suggested that the problem of hierarchical clustering is one of fitting a geometric model—that is, a rooted tree structure—using combinatorial optimization techniques.

In both types of hierarchical procedure the fusions or subdivisions are fixed permanently. Once assigned to a group, the entity or the cluster remains in that group. Unlike iterative partitioning, these methods do not allow reallocation of poorly assigned entities. Another issue concerns the optimum number of clusters present in the data if a hierarchy is not the goal. The agglomerative procedures end in a conjoint cluster containing all entities; the divisive techniques end in a set of N single entities. The problem, then, is how to decide when to stop clustering. Several approaches have been devised to help make this decision (Mojena, 1977; Milligan, 1980). These criteria will be described later.

The Agglomerative Methods

The agglomerative hierarchical methods are the most popular of the clustering techniques. Although the number of different algorithms available is considerable, nearly all are variations of three approaches: linkage methods, centroid methods, and minimum-variance methods. The basic procedure, however, is the same. The process begins with the computation of a distance or a similarity matrix between the $\frac{1}{2}N(N - 1)$ possible pairs of entities. Once the indices are available the matrix is searched for the closest (or most similar) pair i and j. Then i and j are merged to form cluster k and the matrix entry values are modified to reflect the change. The matrix is searched again for the closest pair and the two are merged into a new cluster. The process is followed until all entities are in one cluster. Sneath and Sokal (1973) use the acronym SAHN to characterize the procedure: sequential, agglomerative, hierarchical, and nonoverlapping.

There are several common sorting strategies—that is, ways of defining similarity/dissimilarity between clusters. These clustering techniques are single linkage, complete linkage, average linkage, the

centroid method, the median method, and the minimum-variance method. All these methods of combining clusters have been shown by Lance and Williams (1967) and Wishart (1969a) to satisfy the same recurrence formula. In other words the six methods can be described in terms of the same algorithm. Let the degree of dissimilarity between two clusters i and j be denoted d_{ij}. Suppose the two clusters i and j are merged to form a new cluster k. Then the distance between k and a third cluster h is d_{hk}. The general formula is

$$d_{hk} = Ad_{hi} + Bd_{hj} + Cd_{ij} + D|d_{hi} - d_{hj}| \qquad (6\text{-}1)$$

where A, B, C, D are parameters whose values vary with the cluster method listed in Table 6-1. Assume that the three clusters i, j, and h contain n_i, n_j, and n_h entities respectively with intercluster distances of d_{ij}, d_{hi}, d_{hj}. If the smallest of the distances is d_{ij} and these are combined to form cluster k, then $n_k = n_i + n_j$. The proposed formula is shown in geometric form in Figure 6-1. The diagram shows the relationship between three clusters with respect to two attributes as well as the distances d_{ij}, d_{hi}, and d_{hj}.

As Lance and Williams (1967) point out, the procedure may be characterized as combinatorial and compatible. It is *combinatorial* if the dissimilarity measure d_{hk} can be computed (updated) from the previous measured d_{hi}, d_{hj}, and d_{ij}. The method is *compatible* if the metric among more inclusive clusters is the same as that among the lower-level clusters. As a result the methods that satisfy the recur-

Table 6-1. Parameter Values for the Six Hierarchical Algorithms.

Method	*A*	*B*	*C*	*D*
Single linkage	$\frac{1}{2}$	$\frac{1}{2}$	0	$-\frac{1}{2}$
Complete linkage	$\frac{1}{2}$	$\frac{1}{2}$	0	$\frac{1}{2}$
Average linkage	$\frac{n_i}{n_i + n_j}$	$\frac{n_j}{n_i + n_j}$	0	0
Minimum variance	$\frac{n_h + n_i}{n_h + n_k}$	$\frac{n_h + n_j}{n_h + n_k}$	$\frac{-n_h}{n_h + n_k}$	0
Centroid	$\frac{n_i}{n_i + n_j}$	$\frac{n_j}{n_i + n_j}$	$\frac{-n_i n_j}{(n_i + n_j)^2}$	0
Median	$\frac{1}{2}$	$\frac{1}{2}$	$-\frac{1}{4}$	0

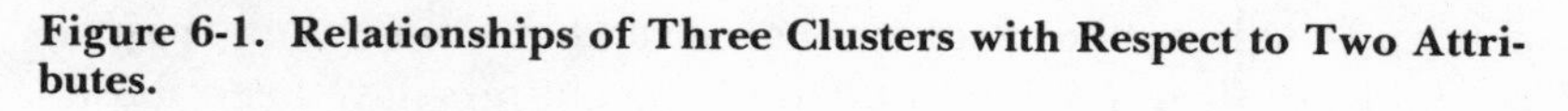

Figure 6-1. Relationships of Three Clusters with Respect to Two Attributes.

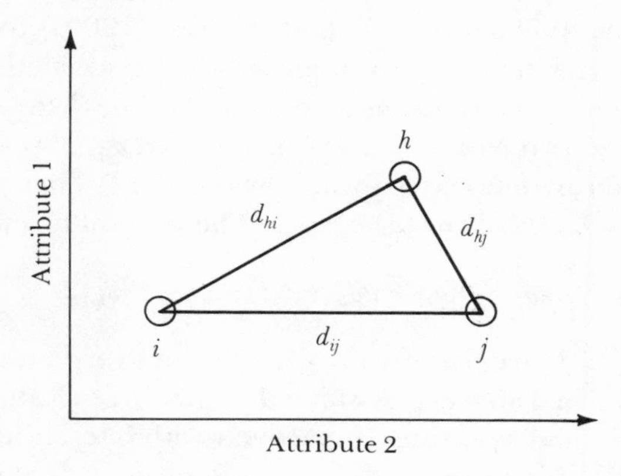

rence formula are computationally rapid and have uniformly increasing (monotonic) distance values. It can be shown that eight well-known hierarchical clustering algorithms satisfy the recurrence formula (Cormack, 1971). As Milligan (1979) indicates, the formula can be used to prove that certain parameter values generate hierarchical structures with monotonically increasing distance values. We shall consider the proofs of such a metric after we have discussed the various sorting strategies.

Single-Linkage Clustering. The method of *single linkage* was introduced independently by Sneath (1957) and McQuitty (1957). It has also been labeled the nearest-neighbor method (Lance and Williams, 1967) and the minimum method (Johnson, 1967). The procedure is sequential, agglomerative, hierarchical, and nonoverlapping.

Initially the N entities are considered as N clusters, each with one member. The first step is to find and merge the two clusters i and j within the set that are separated by the smallest distance d_{ij}. Thereafter the distance between the new cluster k and some other cluster h is defined as the distance between their closest members.

The problem is to find

$$d_{hk} = \min(d_{ik}, d_{jk})$$

where d_{hk} is the distance between the closest members of clusters h and k. Each fusion decreases by 1 the number of groups. If h and k are merged, then the distance between any entity in the resulting cluster and its nearest neighbor is at most d_{hk}. The procedure is called single linkage because clusters are merged by the shortest link between them at each stage. As a consequence every member of the cluster is more like at least *one* other member of the cluster than it is like any other entity not in the cluster.

Should the measures be correlations, the problem is to find

$$r_{hk} = \max(r_{ik}, r_{jk})$$

where r_{hk} is the degree of similarity between the two most similar entities in clusters h and k. Then for any entity in the same cluster at least *one* other entity will correlate at least r_{hk}. As may be seen from Table 6-1, the parameters of the computing formula for single linkage are $A = B = \frac{1}{2}$; $C = 0$; $D = -\frac{1}{2}$. Suppose, for example, $d_{hi} = 3$, $d_{hj} = 4$, and $d_{ij} = 2$. Then $d_{hk} = \frac{1}{2}(3) + \frac{1}{2}(4) - \frac{1}{2}|3 - 4| = 3$.

Complete-Linkage Clustering. The method of *complete linkage* is also called the furthest-neighbor method (Lance and Williams, 1967) and the maximum method (Johnson, 1967). The distance between clusters is now defined as the distance between their most remote pair of entities. The distance between merging clusters is thus the diameter of the smallest sphere that can enclose them. As in single linkage each fusion decreases by one the number of clusters.

The procedure begins with a pair of entities i and j separated by the smallest nonzero value. The distance d_{hk} is then

$$d_{hk} = \max(d_{ik}, d_{jk})$$

where max means the larger of the distances compared and d_{hk} represents the separation between the most remote or distant members of clusters h and k. Should correlation measures be used,

then

$$r_{hk} = \min(r_{ik}, r_{jk})$$

in which min refers to the *lesser* of the two correlations. In the resulting cluster every member is more like *every* other member of the same cluster than it is like any other entity not in the cluster. This means that *all* members of a cluster are linked to each other at some maximum distance or minimum degree of similarity. (In graph-theoretic approaches, complete-linkage clusters are called maximally connected subgroups.) In complete-linkage clustering, the parameters (Table 6-1) are $A = B = \frac{1}{2}$; $C = 0$; and $D = \frac{1}{2}$. Suppose again that $d_{hi} = 3$, $d_{hj} = 4$, and $d_{ij} = 2$. Then $d_{hk} = \frac{1}{2}(3) + \frac{1}{2}(4) + \frac{1}{2}|3 - 4| = 4$.

Average-Linkage Clustering. The method of *average clustering* was first proposed by Sokal and Michener (1958) as a compromise between single linkage and complete linkage. Single linkage leads to long straggly clusters whereas complete-linkage clusters are likely to be small and compact. Sneath and Sokal (1973) differentiate between average and centroid clusterings as well as between weighted and unweighted clustering. In average clustering each member of a cluster has a smaller average dissimilarity with other members of the same cluster than with members of any other cluster. Lance and Williams (1967) call this procedure group-average linkage. Sneath and Sokal (1973) call the process arithmetic-average clustering or UPGMA (unweighted pair-group method using arithmetic averages) and say that among biologists it is probably the most frequently used clustering strategy. The parameters for the method in the recurrence formula (Table 6-1) are as follows:

$$A = n_i/n_i + n_j \qquad B = n_j/n_i + n_j \qquad C = D = 0$$

The distance between clusters is defined as the average of the distance between all pairs of entities in the two clusters. Sokal and Sneath use this distance as a measure of distance between an entity and a cluster. When the method uses correlations, some transform is needed to handle average correlations. One transform is $D^2_{ij} = 2 - 2Q_{ij}$, where Q is the interentity correlation. Lance and

Williams (1967) suggest use of the following formula:

$$d_{ij} = \cos\left[\frac{1}{N_i N_j}\ \Sigma\cos^{-1} Q_{ij}\right]$$

Centroid Cluster Analysis. The method known as *centroid cluster analysis* was originally proposed by Sokal and Michener (1958) and efficiently programmed by King (1967). Sneath and Sokal (1973) call this procedure UPGMC (unweighted pair-group centroid method). Lance and Williams (1967) call it the centroid technique and point out that their combinatorial formula applies only for squared distances. Because the ultrametric inequality is not met, centroid clustering does not yield monotonic hierarchies.

At each stage the two clusters with the closest mean vector or centroid are merged. The new cluster is replaced by the coordinates of the group centroid. Thus the distance between clusters is defined as the distance between the cluster centroids. The centroid itself is a set of variable means across members of the cluster. These cluster members are, on the average, closer to the centroid of their own cluster than to the centroid of any other cluster. The update equation is

$$d_{hk} = \frac{n_i}{n_k}\, d_{hi} + \frac{n_j}{n_k}\, d_{hj} - \frac{n_i n_j}{n_k^2}\, d_{ij}$$

Median Cluster Analysis. When the size of two clusters differs substantially, the centroid of the merged cluster is very close to the larger cluster and thus the characteristics of the smaller groups are lost. Gower (1967) suggested use of the *median* because when i and j are merged the distance of a third cluster h lies along the median of the triangle that joins the three (i, j, and h). (A line called the "median" can be drawn from each vertex of a triangle to the midpoint of the opposite side. The medians intersect in a point.) The distance between clusters is based on the distance between cluster medians. Otherwise the process is the same as the centroid method. The parameters of the iterative formula for this method (Table 6-1) are

$$A = B = \tfrac{1}{2} \qquad C = -\tfrac{1}{4} \qquad D = 0$$

or

$$d_{hk} = \frac{1}{2}d_{hi} + \frac{1}{2}d_{hj} - \frac{1}{4}d_{ij}$$

Flexible Method of Clustering. Lance and Williams (1967) have proposed a procedure that has constraints: $A + B + C = 1$; $A = B$; $C < 1$; $D = 0$. The method was designed to overcome the disadvantage of single linkage—which is chaining—and that of complete linkage—which is too few clusters. They propose use of -0.25 for parameter C. Sneath and Sokal (1973) illustrate the effects of varying the value of C in a series of dendrograms. The procedure of adjusting parameters until the results conform to expectations or preferences can lead to biased findings, however.

Minimum-Variance Method. Ward (1963) and Ward and Hook (1963) have proposed a general hierarchical clustering program. The procedure is based on the premise that the most accurate information is available when each entity constitutes a group. Consequently, as the number of clusters is systematically reduced from k, $k - 1, k - 2, \ldots, 1$, the grouping of increasingly dissimilar entities yields less precise information. At each stage in the procedure the goal is to form a group such that the sum of squared within-group deviations about the group mean of each profile variable is minimized for all profile variables at the same time. The value of the objective function is expressed as the sum of the within-group sum of squares (called the error sum of squares, ESS). Each reduction in groups is achieved by considering all possible $N(N - 1)/2$ pairings and selecting the pairing for which the objective-function value is smallest. Each cluster previously formed is treated as one unit. When the complete hierarchical solution has been obtained, the ESS values may be compared to ascertain the relative homogeneity of the groups formed. A sharp increase in ESS indicates that much of the accuracy has been lost by reducing the number of groups.

Ward's method can be illustrated by applying the algorithm to fighting ship data selected by Cattell and Coulter (1966) from *Jane's Fighting Ships* (1964–1965). Their aim was to find out how many types of ships there were and which ships belonged to each class. The measurements, converted to standard-score form, were as follows: displacement; length; beam; number of light, medium,

heavy, and very heavy guns; number of personnel; maximum speed; continuity of deck construction; submersibility; and number of planes carried. The 29 ships included 5 aircraft carriers, 4 destroyers, 10 submarines, and 10 frigates. Mojena's Rule One (1977) for stopping the clustering process indicated the presence of four groups as anticipated. A large increase in the error sum of squares occurred when three groups were formed. The algorithm correctly recovered all four types of ships without error. A two-dimensional plot showing clusters for 33 ships (not 29) was constructed from an INDSCAL analysis. The groups may be seen in Figure 4-2. It should be noted that two other kinds of ships are included in the scaling study.

Another example of the effectiveness of Ward's procedure comes from a study of body types (Lorr and Fields, 1954). A sample of 90 men was categorized by Sheldon's procedure into three body types. Five men were selected to represent the mesomorphs (muscular), five for the endomorphs (rounded and plump), and five for the ectomorphs (thin and linear). The subjects were intercorrelated on the basis of 36 morphological measures. After the correlations were transformed into D_s^2, Ward's procedure was applied. Mojena's Rule One indicated the presence of three clusters. Figure 6-2 shows the dendrograms of the three clusters in which all members are categorized as predicted.

As mentioned earlier, one defect of most hierarchical procedures is that allocations of entities to clusters are fixed; once assigned, the entities remain in the cluster. As additional entities are allocated the group profile might shift, leaving the original members on the periphery of the group. To correct this deficiency in Ward's clustering process, a two-part computer procedure was developed by Feild and Schoenfeldt (1975). The first part consists of an affirmation program that compares the profile of each entity with the profile of every subgroup and either affirms membership or removes the entity. Removal may occur for several reasons: (1) because the entity is a misfit that should be reclassified; (2) because the entity is an outlier and should not be assigned to any group; or (3) because the entity is an overlap that fits more than one group. The adjusted group means are computed following each change, and the process is repeated until the changes are minimal. The second step uses dis-

Figure 6-2. Dendrograms for Three Body Types from a Minimum Variance Analysis.

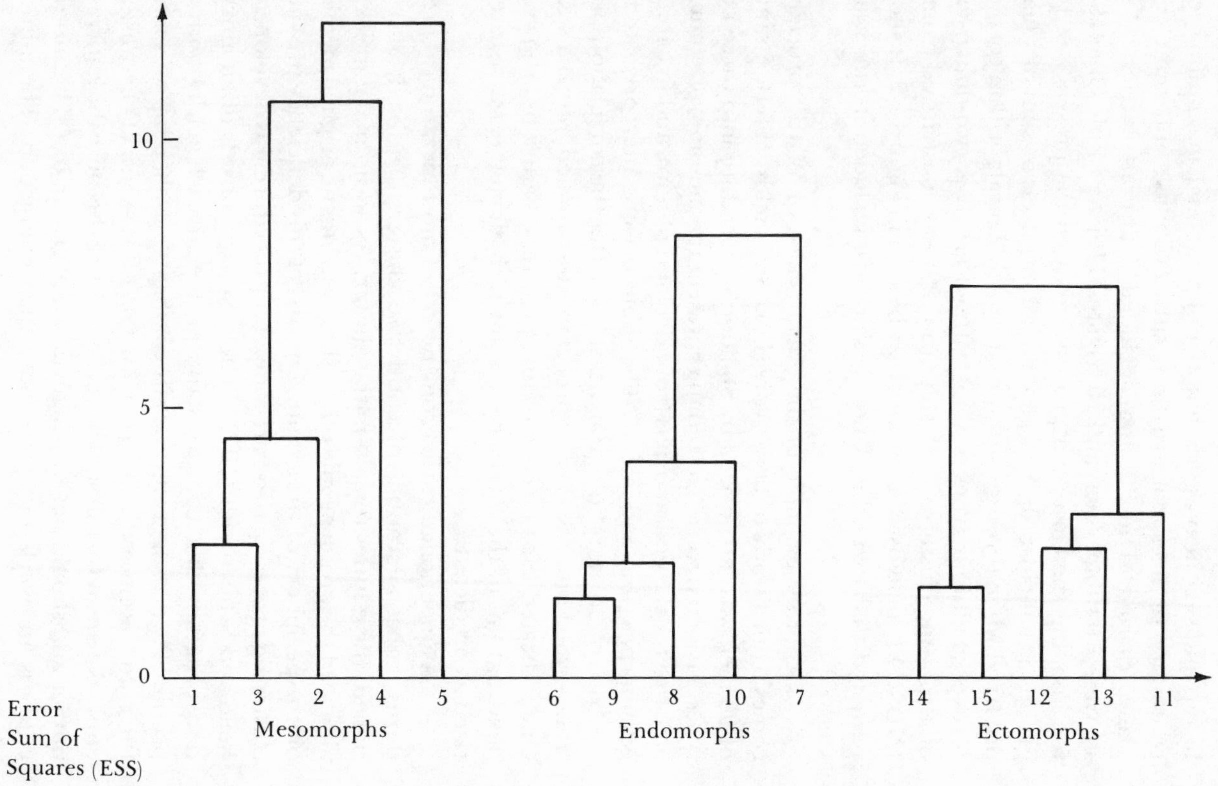

Error
Sum of
Squares (ESS)

10

5

0

1 3 2 4 5
Mesomorphs

6 9 8 10 7
Endomorphs

14 15 12 13 11
Ectomorphs

criminant functions found from the variables and the entities treated as new cases. After each case has been allocated to one of the subgroups, the result is a set of homogeneous clusters.

Ultrametric Hierarchical Schemes

Johnson (1967) has shown that both single linkage and complete linkage induce a metric that satisfies what is known as the *ultrametric inequality.* Recently Milligan (1979), through use of Lance and Williams's recurrence formula, extended Johnson's proof to four other common clustering algorithms: group-average linkage (UPGMA), weighted-average linkage (WPGMA), minimum variance (Ward, 1963), and flexible linkage (Lance and Williams, 1967). Only the centroid and median methods fail to satisfy this inequality. This means that such hierarchical clustering structures have monotonically increasing distance values. If the distances are monotonically increasing, then it is not possible for a cluster merged later in the hierarchy to have a distance value less than the distance value of a cluster merged earlier.

The conditions for a euclidean distance metric were discussed in Chapter Three. In a euclidean space the distances between three points i, h, and j that form a triangle (as in Figure 6-1) must satisfy the triangular inequality

$$d_{(i,j)} \leq d_{(i,h)} + d_{(h,j)}$$

Here $d_{(i,j)}$ is the distance between i and j whereas $d_{(i,h)}$ and $d_{(h,j)}$ are the distances to the third point h. Although the triangle may be of any form, the sum of any two sides must be at least as great as the sum of the third side.

The ultrametric inequality, however, requires that the three distances satisfy the following equation:

$$d_{(i,j)} \leq \max[d_{(i,h)}, d_{(j,h)}]$$

In other words the three distances must form either an equilateral triangle or an isosceles triangle in which the base is shorter than the two equal sides. The ultrametric is clearly more stringent than the

triangular inequality. As Milligan (1980) points out, the advantage of a clustering based on the ultrametric inequality is that even if the measures are subjected to a monotone transformation, it is possible to test how well the data set satisfies the inequality.

The Divisive Methods

A few hierarchical techniques are based on division of the initial data set into subgroups. At each step the objective is to bisect the group into the two subsets that are most distinctive and different; in the agglomerative methods, by contrast, the two entities that are most similar are joined. These divisive methods can be classified as either monothetic or polythetic. *Monothetic* clusters are based on the possession of a single binary attribute that is used to bisect a set of entities. This form of classification is especially useful for the construction of keys or identification lists used in archeology, ecology, and botany. *Polythetic* clusters are based on several attributes shared by members of a cluster. On the whole the agglomerative methods produce polythetic clusters whereas the divisive methods may generate either monothetic or polythetic clusters.

A typical divisive hierarchical process is offered by Edwards and Cavalli-Sforza (1965). Recall that a set of N entities can be divided into two subsets in $2^{N-1} - 1$ ways. The original procedure considered all these possible partitions of N data units into two groups. Their method is to choose the split that minimizes the total error sum of squares for the two groups. This is the same criterion applied in Ward's procedure. The method is obviously limited to very small problems because of the great number of possible partitions. Computer programs for this monothetic method may be found in the DIVIDE algorithm of CLUSTAN and in Hartigan's book (1975).

McNaughton-Smith and others (1964) have proposed a polythetic divisive technique called dissimilarity analysis. The measure of dissimilarity applied is the average euclidean distance between each entity and all others in the set. The entity used to initiate a subgroup is the one most distant from all others. Next the average distance of each entity to the new subgroup and the average dis-

tances of each entity to the main group are determined. The difference between these two averages is then found. The entity most distant is assigned to the new subgroup. The process is repeated until all differences are negative. Subsequent steps may be continued to divide the subgroups into further subgroups.

Blockmodels

The idea of permuting discrete or binary data matrices in order to reveal blocks or patterns was the original motivation behind blockmodels. A blockmodel is a binary data matrix so arranged as to reveal blocks or patterns of high and low density values (1s and 0s) for a set of attributes. Perhaps the earliest representation of structure in data by blocking was given by Lambert and Williams (1962) in listing plant species by site. The goal was to permute the $N \times m$ matrix so that the rows and columns of 1's and 0's could be partitioned into homogeneous groups. In sociology one or more social ties such as likes or dislikes that yield an $N \times N$ matrix can be permuted into blocks of high density (*one blocks*) and low density (*zero blocks*). By a *block* is meant a subset of rows and columns of a data matrix on which a blockmodel has been imposed. The density for a binary matrix is the number of units divided by the number of entries. The term *image* refers to a matrix in which the entries in a submatrix have been replaced by either 1 or 0. A 1 corresponds to a block containing at least some 1's; a zero in the image corresponds to a zero block in the matrix.

At present the most widely used algorithm for obtaining blockmodels and image matrices is called CONCOR (convergence of iterated correlations). Its formal properties have been reported by Breiger, Boorman, and Arabie (1975) and by Arabie, Boorman, and Levitt (1978). McQuitty and Clark (1968) independently devised the same algorithm and labeled it IICA (*iterative-intercolumnar correlation analysis*). CONCOR (or IICA) is a divisive procedure. The method can begin with a matrix of interassociations between N entities and values of 1 in the diagonal cells of the matrix. Or it can begin by forming a new square matrix of product-moment correlations between columns of the original $N \times m$ data matrix. The correlations

between the corresponding entries of any two columns are computed and assembled in an intercolumnar matrix. A second intercolumnar correlation matrix is then computed following the same procedure. Iterative application of this procedure will in general converge to a matrix in which all coefficients are either +1 or −1. The final matrix may be permuted into a bipartite (two-block) form:

$$\mathbf{M} = \begin{bmatrix} +1 & -1 \\ -1 & +1 \end{bmatrix} \tag{6-2}$$

Each submatrix can be partitioned into two and the process continued to produce a hierarchical classification of the original set from the top down. Convergence, according to the authors, is usually rapid.

Form of Input Data. Most applications of CONCOR enter integer values that are either binary or ordered choices. The binary form usually provides the clearest data for interpretation to blockmodels. The principal advantage (Arabie and Boorman, 1982) of binary data once a blockmodel has been obtained is that such data sharpen the pattern of contrasting densities between potential zero blocks and one blocks. Multiple types of social networks that exhibit maximal relational contrast—symmetric versus asymmetric ties, for example, or positive and negative affect—provide the most interesting interpretations.

Applications. The value of CONCOR and blockmodels will be illustrated by two studies. The first is by Breiger, Boorman, and Arabie (1975) who analyzed data collected by Sampson (1969). The second was reported by Ennis (1982). Sampson has given a careful account of social relations in an isolated American monastery in the late 1960s. There was a major conflict in the monastery that led to a mass departure of members. One block of data included answers to sociometric questions regarding affect, esteem, influence, and sanctioning. Each respondent was asked to list in order, for example, the three brothers he esteemed most and least. Responses for 18 members were reported for five time periods. After CONCOR was applied to these weighted choices, the partitioning process resulted in three blocks: Loyal Opposition, Young Turks, and Outcasts. Figure 6-3 shows how the total set of 18 was subdivided into two. Then each subset in turn was subdivided into two blocks. The

Figure 6-3. Hierarchical Clustering Representation of Repeated Application of the CONCOR Algorithm on Sampson's Data.

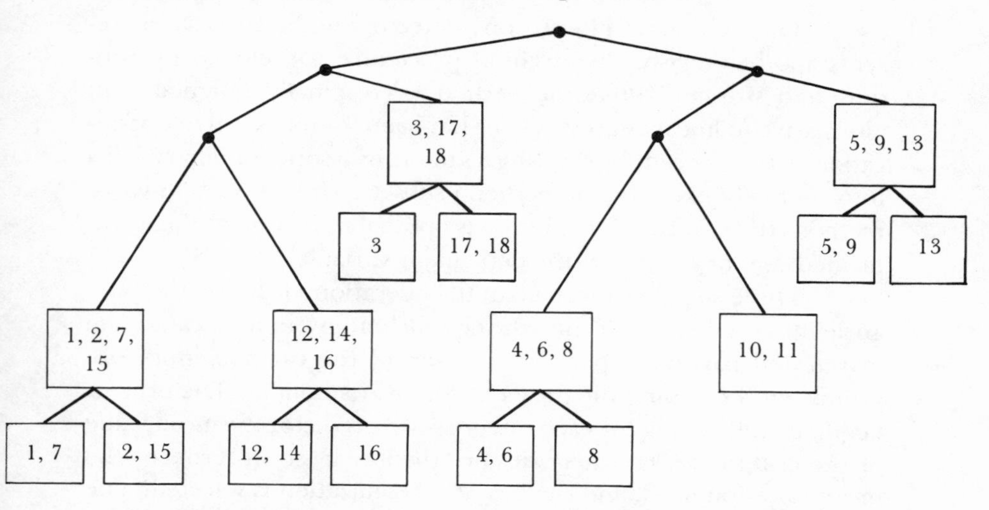

Source: Brieger, Boorman, and Arabie, 1975. Used by permission.

process evidently was stopped when the blocks were too small or uninterpretable.

Another application is reported by Ennis (1982). Bales, Cohen, and Williamson (1979) had developed a set of dimensions to characterize group behavior as measured by SYMLOG (systematic multiple-level observation of groups). The three dimensions refer to dominance versus submission, friendliness versus hostility, and task orientation and instrumental behavior versus emotional and expressive behavior. Bales and his colleagues collected data on a group of 12 members and a leader. During the fourth and seventh week, members rated each other on the SYMLOG adjective rating form. The CONCOR algorithm was applied to the 13 × 13 matrices, and these were partitioned into three blocks. Members of block 1, called Leaders, were most friendly, outgoing, and dominant. The Onlookers of block 2 were more withdrawn, silent, and less participant. The third block, called Rebels, were most frequently engaged in arguments concerning leadership and power. Application of multidimensional scaling (KYST) to the same data yielded two dimensions that confirmed the findings of CONCOR blocking.

The AID Method

The automatic interactions detector method (AID) represents another divisive hierarchical procedure. Developed by Sonquist and Morgan (1964), the method was originally designed as an alternative to linear regression but has been found useful in classification. AID proceeds by dividing a sample by a series of binary splits into mutually exclusive monothetic subsets. Unlike other divisive methods the criterion for splitting is optimal reduction in the unexplained sum of squares of the dependent variable.

Three steps are involved in the operation: ordering the categories in a predictor, selecting the best dichotomy, and repeating the search and partition operation in each of the two resulting subgroups until reaching the preset criteria for stopping. The process begins by calculating for each category of a predictor the mean value of the criterion. The categories are then ordered in terms of the mean criterion level, and the best dichotomization is selected. The best split is the one that maximally reduces residual or unexplained variation. This residual variation, the sum of squared deviations around the subgroup means (WSS), is zero only if all observations in the dependent variable are the same in each subgroup. The total sum of squares (TSS) is the sum of squared deviations around the mean of the total group. The difference between TSS and WSS is the between-group sum of squares (BSS). Hence minimizing residual variation (WSS) is the same as maximizing BSS/TSS. If the subgroup sizes are N_1 and N_2 and the criterion means are $\bar{X}_1$ and $\bar{X}_2$, the value of BSS is as follows:

$$\text{BSS} = \frac{N_1 N_2}{N} (\bar{X}_1 - \bar{X}_2)^2$$

The AID method offers a clustering alternative to linear regression. It may be used both with discrete and with continuous predictor variables. Illustrations and computer programs may be found in Hartigan (1975) and the SPSS statistical package (Nie and others, 1975). Useful illustrations concerning drug research and college dropouts may be found in Somers and others (1976).

Stopping Rules

A primary objective in many studies, especially in biology, is to generate a hierarchical structure. In the behavioral and social sciences, however, the goal is usually to find natural groups and to reproduce underlying structure. Suppose underlying structure is viewed as consisting of mixtures of samples from multivariate populations. Then if a hierarchical procedure is applied one must determine the level that best reproduces the structure. In short, a *stopping rule* is required to select the optimum number of clusters.

Mojena's Rule. Mojena (1977) developed two structural stopping rules and evaluated seven hierarchical methods described earlier in the chapter. The computer was used to simulate 12 samples of size 30 based on five orthogonal variates. The stopping rules were evaluated for Ward's method with the index (alpha) defined in terms of standardized euclidean distance. Only Rule One proved effective.

Mojena's Rule One uses the distribution of the clustering criterion (the within-group sum of squares) to determine when a "significant change from one stage to the next implies a partition which should not be undertaken" (Mojena, 1977, p. 359). Three values are needed in the formula: the mean $\overline{X}$ of the distribution of the clustering criteria; the unbiased standard deviation S of their distribution; and the standard deviate k (z score for significance level). In terms of the predicted number of clusters, Mojena found that values of k in the range of 2.75 and 3.50 gave the best overall results. More (1981) found that values of 3.0 and 3.5 worked equally well. The formula for Rule One, then, is as follows:

$$\text{Alpha} = \overline{X} + kS \qquad (6\text{-}3)$$

With appropriate adaptation the rule could be applied to measures of the error sum of squares derived for each of the various linkage methods.

Milligan's Null-Hypothesis Test. Another approach is to test the null hypothesis that the observed sample of clusters was obtained from random noise data lacking in cluster structure. Milligan (1980) suggests that a point-biserial correlation be used to measure the

goodness of fit of the input data to the resulting cluster partition. The correlation is computed between elements of the input distance matrix and a corresponding matrix consisting of 0 and 1 entries. A value of zero is assigned if two corresponding entities are clustered together by the algorithm; a value of 1 is assigned otherwise. The test is conducted by generating random noise data sets. The random data sets are then analyzed by the clustering algorithm. The value of the point-biserial correlation is then determined at each successive partition level. If this value, obtained from a clustering of an empirical data set, is significantly deviant with respect to the values generated by the random data sets, the null hypothesis can be rejected. A significant point biserial correlation implies that the partition level represents a nonrandom set of clusters.

In agglomerative hierarchical procedures the N entities are successively merged into fewer and fewer clusters until the conjoint set contains all entities. In divisive hierarchical methods the conjoint set is split successively into smaller and smaller subsets until there are N clusters, each containing one entity. Thus the problem of choosing a level that will provide an optimum number of clusters for the data is the same for both types of methods. As yet, however, no acceptable stopping rules have been developed for the divisive procedures with the exception of AID.

An Evaluation

A fair number of hierarchical cluster programs have been reviewed in the chapter. The question remains, however, as to which algorithms are the most effective in recovering an underlying structure. Another hidden problem is that any such procedure will generate a clustering of any set of data, even with random noise data lacking structure. For this reason an increasing number of studies are being conducted to evaluate the various methods. A few will be reviewed here; a more thorough analysis of empirical studies is presented in Chapter Seven.

Sneath (1966) and Cunningham and Ogilvie (1972) found average linkage and complete linkage superior to single linkage. Milligan (1980) likewise found that average linkage resulted in the

greatest recovery (complete linkage next) compared to minimum variance and single linkage. Blashfield (1976b), using clusters generated by Monte Carlo methods as Milligan did, found minimum variance best; Kuiper and Fisher (1975) too found Ward's method best. Later Blashfield and Morey (1980) generated three MMPI clusters by Monte Carlo techniques; here group-average linkage proved more accurate than Ward's minimum-variance procedure.

On the other hand, Bartko, Strauss, and Carpenter (1971) found that complete linkage best reproduced clusterings of archetypal psychiatric patient classes. Mezzich (1982) conducted a comprehensive comparison of 10 taxonomic methods using four data bases including the archetypal psychiatric categories. Complete linkage and centroid linkage consistently ranked highest among the hierarchical clustering techniques.

At present, group-average linkage appears best and somewhat superior to the minimum-variance method. The latter is especially sensitive to elevation (mean level) in the data and performs best where clusters are well separated. Complete linkage does best when tests are run on known cluster structures as in Mezzich and in Bartko and colleagues. Single linkage, however, is still needed to detect elongated or curved cluster structures. It is most affected by outliers and intermediates. Because of the state of the art, the researcher using clusters should apply at least two methods to confirm that the underlying cluster structure is being recovered. Another necessity is replication on several sets of data.

Cluster Analysis Programs

Blashfield (1976b) and Blashfield and Aldenderfer (1978) have published several reports on cluster analysis software packages. They found three basic sources of clustering programs: general statistical packages, programs that perform only one type of cluster analysis, and programs collected in books. The general statistical packages are CLUSTAN, NTSYS, BMDP, and OSIRIS. The programs that concern a single method are HGROUP (Veldman, 1967) for minimum variance and HICLUS (Johnson, 1967) for the single-linkage and complete-linkage methods designed by Johnson. Nearly

all the principal hierarchical clustering programs (single, complete, and average linkage as well as Ward's) may be found in Anderberg (1973), Hartigan (1975), Wishart (1978), and Rohlf, Kishpangh, and Kirk (1971). Details concerning programs are given in Appendix A at the back of the book. Aldenderfer and Blashfield (1978) offer a useful survey of computer programs for performing hierarchical cluster analysis.

Summary

The hierarchical clustering techniques are classifiable into the agglomerative and the divisive. The agglomerative begins with N entities which are sequentially merged at successive levels until all are included. The divisive procedures partition the conjoint set into two subsets and each subset is partitioned into still finer subsets. The agglomerative methods are separable into the linkage methods, the centroid methods, and minimum variance. The linkage techniques include single, complete, and average linkage. Eight of these clustering algorithms satisfy a recurrence formula developed by Lance and Williams that simplifies programming. Each of these methods is defined and assessed for positive and negative features. Several examples are given (fighting ships and body types) to illustrate minimum variance. It is shown that eight of the linkage methods induce a metric that satisfies the ultrametric inequality and hence the measures are invariant to monotone transformations. Several stopping rules or procedures for determining the optimum number of clusters to retain when agglomerative hierarchical procedures are applied, are examined. Next the divisive techniques are differentiated into those that generate monothetic (based on a single attribute) clusters and those that develop polythetic clusters (based on several attributes). Three divisive methods including a new procedure called blockmodeling are described. Finally a method called AID is presented. It builds subgroups on the basis of predictors that best explain variation in the dependent variable. The chapter concludes with a brief evaluation of relative effectiveness of the various hierarchical cluster procedures. Group average linkage appears best

and somewhat superior to minimum variance. Single linkage is, however, still needed to detect straggly, elongated clusters.

In the next chapter we turn to empirical studies that seek to evaluate, usually by Monte Carlo procedures, the accuracy of the clustering solutions obtained. Some internal and external measures of goodness of fit will be suggested.

CHAPTER 7

Empirical Studies of the Cluster Methods

Research concerned with the accuracy and stability of clustering methods has grown at a rapid rate. This research is of considerable value because evidence is needed that clusters correspond to the populations from which the data samples were drawn. The fundamental question is this: Which clustering algorithm can recover the true cluster structure? Some of the methods impose cluster structure rather than find it. As Sneath and Sokal observe (1973, p. 252): "Cluster methods will yield clusters of some kind whatever the structure of the data, even if the distributions are random." This assertion is clearly correct when agglomerative hierarchical cluster methods are applied, for at each level the two closest clusters are combined. Moreover, when the data do not satisfy the assumptions of a particular technique, the method may impose a nonexisting structure on them. One would also like to know which methods are robust to recovery when the structure is obscured by error or noise.

Several views of cluster structure were examined in Chapter Five. One model, now widely adopted, contends that clusters represent mixtures of samples from multivariate normal populations (Rand, 1971; Kuiper and Fisher, 1975; Blashfield, 1976a; Edelbrock, 1979). A second model is primarily spatial. Everitt (1974), for example, generates two-dimensional data sets that can be examined

visually. The concept of ultrametric space as the basis for the generation of clusters represents a third model endorsed by Jardine and Sibson (1971) and Johnson (1967). A fourth approach to testing cluster structure is to create data sets based on known classes. Cattell and Coulter (1966) have used classes of fighting ships, dogs, and ethnic groups. Helmstadter (1957) used geometric figures (spheres, cylinders, tetrahedrons); Lorr and others (1955) used pyramids, cylinders, prisms, and cones. Bartko, Strauss, and Carpenter (1971) generated data sets consisting of factitious psychiatric patients with distinctive symptom patterns.

This chapter reviews the major published empirical comparisons of the most frequently used clustering techniques. Most of the studies used computer-generated clusters by means of what are called Monte Carlo methods. These methods will be discussed briefly. We shall also consider measures of accuracy of recovery. These measures reflect the degree of agreement between obtained clusters and known underlying populations. The chapter concludes with a summary of the findings regarding the various clustering techniques.

Recovery Measures

Three types of measures have been proposed to assess the validity of a data partition. The first represents an *external criterion* that uses information obtained outside the clustering process to evaluate cluster recovery. This approach involves estimating how accurately the cluster solution matches the true structure of the data set. The second type represents an *internal criterion* that uses information from within the clustering process. Such measures reflect the goodness of fit between the input data and the resulting data partition. In other words: How well does the partition approximate the similarities of the entities? A third approach for evaluating a cluster structure is to estimate the *replicability* of a cluster solution across successive data sets. If a cluster structure is repeatedly discovered within different samples from the same populations, the solution is likely to have some generality. This method is especially appropriate if the true structure is not known—which is generally the case in applied research. Accuracy and replicability, however, are not neces-

sarily highly correlated. Although replicability is necessary for accuracy, accuracy is not essential to replicability. In the following sections we shall examine these three types of measures in greater detail.

External Criteria. The two most popular external criteria receive our attention here. A more thorough discussion is available in reports by Rohlf (1974) and Milligan (1979). The two most frequently applied measures are Rand's (1971) index and Cohen's statistic kappa (Cohen, 1960; Fleiss and Zubin, 1969).

Rand's index has fixed limits of 0.0 to 1.0, where 1.0 represents perfect agreement. The index is quite general since it represents the degree of recovery for a single partition. As such it may be applied both to hierarchical and nonhierarchical algorithms. Consider a square matrix whose entities S_{ij} are either 0 or 1. Now S_{ij} equals 1 if entities i and j are clustered together in both the criterion cluster and the obtained cluster; S_{ij} also equals 1 if entities i and j are not clustered together because the pairs come from different populations and are placed apart in the obtained clusters. The entry is given a value of zero if in one cluster the pair are together and in the partition the pair are not together. The statistic (Edelbrock, 1979) is

$$\frac{m_1 + m_2}{N(N-1)/2} = R$$

where m_1 is the number of entity pairs from the same population placed together in the obtained clusters, m_2 is the number of entity pairs from a different population placed apart in the obtained clusters, and N is the number of entities. For hierarchical methods the index is applied at the level of the hierarchy corresponding to the exact number of clusters known to be present in the data. The nonhierarchical programs can be preset also to recover the correct number of clusters.

Another index of the adequacy of cluster recovery is kappa, which is defined as

$$\text{kappa} = \frac{P_o - P_c}{1 - P_c}$$

where P_o is the observed proportion of classification agreement and P_c is the proportion of agreement expected by chance. The statistic value of 1.00 indicates perfect agreement, but there is no fixed lower bound. To the extent to which nonchance factors are operating, P_o will exceed P_c. The difference represents the proportion of cases in which beyond-chance agreement occurred. Thus kappa is the proportion of agreement after chance agreement is removed from consideration. Kappa can be used, in the context of cluster analysis, only at a specific level of the hierarchy associated with the known or correct number of clusters. Hence it cannot be used to measure partial recovery of the clusters. A further problem is that it does not produce unique values. The number of possible values increases with the number of clusters. Blashfield (1976b) used the highest numerical value as the recovery measure for a given set. For these reasons Rand's index may be preferred over Cohen's kappa. Both Edelbrock (1979) and Milligan and Isaac (1980) report that Rand's index and kappa correlate above 0.97.

Internal Criteria. A variety of goodness-of-fit measures have been proposed by Hubert and Levin (1976). Milligan (1980) has suggested a simple point-biserial correlation between the raw input dissimilarity matrix and a corresponding matrix consisting of 0 and 1 entries. A value of zero is assigned if the two corresponding entries are clustered together by the algorithms; a value of 1 is assigned otherwise. The formula for a point-biserial coefficient of correlation reads

$$r_{bi} = \frac{M_p - M_q}{\mathrm{SD}} (pq)^{\frac{1}{2}}$$

Here M_p is the mean dissimilarity of the higher group in the dichotomized variable and M_q is the mean of the lower group; p is the proportion of cases in the higher group, q is the proportion in the lower group, and SD is the standard deviation of the total sample. A point-biserial correlation (Guilford, 1965) represents the correlation between a continuous variable (the dissimilarities) and a dichotomous variable (the 0 and 1 values).

Another useful criterion measure is called gamma (W. H.

Kruskal, 1958). It was found to be the best of 30 criteria examined by Milligan (1981a). This measure of association is appropriate for ordered contingency tables.

$$\text{Gamma} = \frac{ad - bc}{ad + bc} = \frac{P - Q}{P + Q}$$

The first formula presents the frequencies taken from an ordinary 2×2 table. In the second formula P is the number of concordant pairs whereas Q is the number of discordant pairs. Gamma may take any value in the range -1 to $+1$.

Milligan (1980) conducted a computer-generated study of 30 internal indices. The best four internal indices were gamma, the C index (Hubert and Levin, 1976), the point biserial (Milligan, 1980), and tau (Rohlf, 1974).

A Measure of Replicability. McIntyre and Blashfield (1980) developed a procedure called the *nearest-centroid technique* for evaluating Ward's method. Since stability (replicability) is a necessary condition for accuracy and can, moreover, be measured directly, it offers a basis for indirectly estimating the accuracy of a solution. Let us consider the cross-validation paradigm used in *multiple regression.* The stability of the parameters is tested by deriving the regression coefficients in the derivation sample and by applying these coefficients to another random or *holdout* sample. Predicted values of the criterion in the holdout sample are obtained by applying the derivation sample's linear coefficients to the new data. The actual values of the criterion in the holdout sample are then correlated with the predicted values. The difference between the multiple correlation in the holdout sample and the cross-validated multiple correlation provides an estimate of the stability of the regression coefficients.

Since cluster analysis has no linear coefficients that can be applied to multiple random samples, one must resort to the assignment rule applied in *k*-means partitioning. When two samples are involved, the entities in the second data set are assigned to one of the centroid vectors of the first data set on the basis of the smallest euclidean distance. The degree of agreement between the nearest-centroid assignment and the results of the cluster analysis of the second sample represents an estimate of the cluster solution's repli-

cability. The steps in this evaluation procedure are as follows: (1) Select two independent random samples A and B from a data set; (2) cluster-analyze sample A; (3) compute the centroid vectors for each cluster; (4) cluster-analyze sample B; (5) compute the squared distance of each member of B from the centroids of A; (6) assign each member of sample B to the nearest vector; (7) apply kappa to measure the degree of agreement between the members of step 6 and the cluster members of step 4. The index is called *agreement kappa.*

Two studies were conducted to evaluate this hypothesis. In the first study the procedure was tried with computer-generated multivariate normal data sets. The minimum-variance procedure was used for clustering; three populations were assumed; the major factor varied was cluster overlap. The Spearman correlation between accuracy kappa and agreement kappa was 0.89, suggesting that the stability measure (agreement kappa) was predictive of the accuracy of the cluster solution. In the second study three factors were varied: skewedness of the marginal distribution of the variables, degree of variable intercorrelation, and degree of population overlap. When there is little overlap, agreement kappa is high; with much overlap, kappa is low. Agreement kappa is moderately to highly correlated with accuracy kappa, implying that it does provide an indirect estimate of the accuracy of minimum-variance solutions.

Monte Carlo Procedures

In many of the studies to be described there is frequent reference to *Monte Carlo studies*—that is, studies in which statisticians use a random number generator as a source of data for examining the operating characteristics of their statistics. The properties of complex statistical procedures can often be discovered by empirical methods. Random numbers are used to construct hypothetical data that will fit a statistical model. The statistical method is then studied by applying it to successive data sets. Computations based on random numbers are referred to as Monte Carlo computations.

Although the customary source of random numbers is printed tables, storing extensive tables of numbers in the computer is

costly and inconvenient. However, there are several mathematical formulas that can produce sequences of numbers that behave as if they were random (Green, 1963). The numbers are called *pseudo-random* because they derive from completely determined calculations, but they satisfy all the statistical tests of randomness. Two different approaches are used for generating random numbers by a computational process. In both approaches the process is sequential in that each number is derived from the number or numbers preceding it. In one case the process involves addition; in the other, multiplication is used. Both procedures can generate long sets of digits that pass the statistical test of randomness.

Reviewing Monte Carlo Studies

Recently Milligan (1981b) reported on a thorough review of Monte Carlo validation studies of clustering algorithms. He divided this literature into three periods: the first dating up to 1975, covering early Monte Carlo studies; the second, covering the years 1975 to 1978; and the third and most recent period, including 1979 and 1981. Some of these studies and their findings are reviewed in the sections that follow.

Everitt. Everitt (1974) conducted a series of studies using two-dimensional data sets. The data sets were constructed by generating random variables from bivariate normal distributions. The first data set consisted of a sample from a single group; the rest were constructed to contain two groups. In some cases equal samples of 100 were taken from each population; in other cases, unequal sample sizes of 100 and 20 were used. Some data included elliptical clusters whereas others were spherical. The last data sets, each of 100, defined spherical and elliptical clusters.

Among the clustering methods tested were several nonhierarchical optimization techniques, three hierarchical techniques, and two density or mode-seeking techniques. The two optimization criteria applied were taken from McRae (1971), Friedman and Rubin (1967), and one suggested by Marriot (1971). One index that minimized the within-group sums of squares actually *imposed* a circular group structure on the two elliptical clusters that were present.

Half the points in each elliptical cluster were identified as members of one cluster and the other half as members of the second cluster. On the other hand, Friedman and Rubin's $|\mathbf{W}|$ clustering criterion did recover the two elliptical groups generated. Evidently some techniques may impose structure on data.

The three hierarchical techniques applied were single linkage, centroid linkage, and minimum variance. When they were applied to 50 points representing a *single* cluster, their dendrograms were markedly different. The single-linkage dendrogram indicated no structure, but dendrograms for the other two methods suggested the presence of two groups. When the methods were applied to two populations, each indicated the presence of two groups. The three techniques were then applied to two well-separated elliptical groups. Single linkage recovered the two groups; the other two methods found two groups, but they were not those originally generated.

When NORMAP and NORMIX, the density-seeking techniques, were applied to a single distribution, only a single group was found. The two circular groups were recovered, but the significance tests were doubtful. NORMAP successfully recovered the two elliptical groups.

Everitt concludes that many cluster methods are biased toward finding circular or spherical clusters. A classic example of data containing nonspherical clusters is the Hertsprung–Russel diagram, which shows the plot of stellar temperature versus luminosity. Two elliptical clusters of stars are clearly visible, but many clustering techniques (such as Ward's) fail to find the correct solution (Forgy, 1965).

Cunningham and Ogilvie. Cunningham and Ogilvie (1972) compared seven common hierarchical techniques including single linkage, complete linkage, group-average linkage, centroid linkage, median linkage, and Ward's sum of squares method. The input data consisted of six sets of 20 objects. The first set was composed of four well-separated clusters; the second contained fairly close clusters; the third comprised 20 uniform points; the fourth consisted of 20 chained points. Two goodness-of-fit measures were chosen to assess the grouping methods. One was the rank correlation (tau) between d_{ij} and d^*_{ij}, the input and output distances between object pairs. The

second was a stress measure:

$$\frac{\Sigma(d_{ij} - d^*_{ij})}{\Sigma d^2_{ij}}$$

The group-average methods proved to be most accurate.

Blashfield. Blashfield has reported two studies. In his first study (1976a) 50 multivariate normal data sets with known structure were generated using Monte Carlo techniques. Each data set was a mixture of samples from a number of different populations. The parameters of the mixture reflected the number of populations, the number of variables, the number of entities in the sample, the number of principal components, and so forth. The methods compared were single linkage, complete linkage, average linkage, and minimum variance. The statistic kappa was used to measure the accuracy of the cluster solutions. The minimum-variance method generally formed the solutions that had the greatest accuracy in recovering the mixture structure. Single linkage resulted in the poorest classification; average linkage ranked next; complete linkage obtained the second most accurate solutions.

In the second study (Blashfield and Morey, 1980) four clustering methods were compared using Monte Carlo procedures to generate data sets that resembled MMPI psychotic (8-6), neurotic (1-3-2), and personality disorder (4-9) patterns. The methods compared were inverse factor analysis, Ward's method, average linkage, and Lorr's nonhierarchical average linkage. The model was used to create vectors of means, standard deviations, and scale intercorrelations that resembled real data. The parent sample included three groups of 30 profiles. The criteria applied were the number of misclassifications and coverage (defined as the number of profiles not assigned to any cluster). Hierarchical average linkage with Pearson correlations yielded three clusters with 91 percent coverage and three misclassifications. Replication yielded 100 percent coverage and 12 misclassifications. Ward's method proved to be most sensitive to profile elevation since the clusters tended to be based on elevation. Ward's coverage was automatically 100 percent. Transpose factor analysis resulted in three factors with 80 to 84 percent coverage and 8 to 12 misclassifications. Nonhierarchical average linkage

was rerun after an error was discovered (Lorr, 1981; Blashfield, 1981). The method yielded results that varied with inclusion-exclusion cutoffs. In the best run there were no misclassifications but the coverage was small.

Kuiper and Fisher. Kuiper and Fisher (1975) conducted a Monte Carlo comparison of six clustering procedures: single linkage, complete linkage, median linkage, group average, centroid linkage, and Ward's sum of squares. Their aim was to compare the methods on bivariate and multivariate normal Monte Carlo samples. These hierarchical procedures were stopped for the correct number of clusters and compared for percentage of correct classification. Rand's index was used to evaluate the accuracy of the results. Ward's sum of squares was best, and complete linkage ranked next for clusters of equal size. Single linkage was the poorest of the methods: the presence of outliers usually created problems.

Edelbrock. Edelbrock (1979) observed one major limitation of mixture-model studies of agglomerative clustering techniques. In such studies all objects must be classified to calculate the accuracy or completeness of the recovery. Due to the influence of outliers (entities that belong to no group) the tests severely underestimate the accuracy of the algorithm. What is needed is a technique for evaluating accuracy that does not require 100 percent coverage but considers accuracy as a function of the coverage achieved by a subset of clusters. The approach taken by Edelbrock was to calculate accuracy at the level where the number of actual clusters first equaled the number of underlying populations and then at successively lower levels until accuracy equaled 1.00. The statistic kappa was used to measure accuracy.

Ten multinormal mixtures were selected from 50 mixtures generated by Blashfield (1976a) using Monte Carlo techniques. The clustering algorithms compared were single linkage, complete linkage, average linkage, centroid linkage, and minimum-variance technique. To study the influence of the index of similarity, the algorithms were applied both to product-moment correlations and to euclidean distances. The minimum-variance technique was applied only with distances. To assess bias, a hierarchical clustering algorithm that chooses amalgamations randomly was developed and applied to the same data sets; accuracy was calculated in the same

manner. To evaluate the value of standardized scores, the accuracies of clustering solutions of standardized and unstandardized versions of the same data were compared.

The results revealed that higher accuracies were obtained for standardized data sets than for unstandardized data sets at high levels of coverage. At lower levels there was no difference. Edelbrock also found that average linkage was significantly more accurate than the other procedures with both correlations and distances. Moreover, algorithms using correlations as measures of similarity yielded more accurate recovery rates than distances. All the algorithms were significantly more accurate than the random-linkage algorithm. In summary, the argument is that in most applications of cluster analysis in the social sciences the objective is simply to identify homogeneous subgroups that share a common set of characteristics. Cluster analysis is used as a heuristic tool for creating one of several possible empirical classifications. An algorithm that classifies fewer entities more accurately while providing a distinctive profile may be more desirable than one that classifies everyone.

Bayne. Bayne and colleagues (1980) have reported on Monte Carlo comparisons of selected clustering procedures. Their goal was to use Monte Carlo simulations to estimate the probability of misclassification of 13 clustering methods. Six parameters were varied for the two bivariate normal populations. The nine agglomerative hierarchical techniques included were single linkage, complete linkage, centroid and median linkage, group average, weighted average, Ward's sum of squares, and two other methods. The two partitioning procedures chosen were convergent k means and the trace of $\mathbf{W}$ (within-group sum of squares). Wolfe's maximum-likelihood procedure represented the density-seeking approach. Finally a discriminant function was applied to the known clusters.

All the clustering methods compared favorably with the discriminant-function results except for single linkage. (A discriminant function assumes that the number of groups is known.) The group-average procedure was among the best for large-distance separations of the populations. The highest percentage of misclassification was for Wolfe's maximum-likelihood procedure (see Chapter Five).

In general, the greater the distance between populations the smaller the percentage of misclassification. The best three procedures proved to be k means, Ward's sum of squares, and the trace of **W**. Another important finding was that high variable intercorrelations and differences in population size adversely influence all methods.

Milligan and Isaac. A simulation study was conducted by Milligan and Isaac (1980) to evaluate the ability of four hierarchical clustering algorithms to recover the true structure in data sets. These four procedures were single linkage, complete linkage, average linkage, and the minimum-variance method. All four algorithms satisfy the ultrametric inequality and hence have produced hierarchies of distances that increase monotonically. The cluster structures were based on three factors: type of structure, maximum within-cluster distance, and error conditions (seven levels). The definition of cluster structure was based on external isolation and internal cohesion.

The type of cluster structure was varied systematically. Included were a random set, two cluster sets varying in number of points, and three cluster sets varying in size. The clusters varied in separation distance relative to within-cluster distance. Error was systematically varied with distance. Cluster recovery was measured by kappa and Rand's index. The Spearman correlation was computed between each error-perturbed matrix and the pure ultrametric matrix from which it was derived.

Of the 630 matrices generated, the group-average method had the highest recovery rate and found the correct partition in 442 data sets (70 percent). The complete-linkage method placed second with an overall recovery rate of 402 (64 percent). It was also more sensitive to error perturbation than the group-average method. Ward's minimum-variance method placed third with a recovery rate of 358 (57 percent). In the zero-error-level condition, however, it recovered 10 percent of the data sets, which implies that the minimum-variance criterion was inappropriate to the data. Single linkage, with a recovery rate of 191 sets (30 percent), placed last. The Rand and kappa indices indicated that single linkage was functioning at a chance level in regard to correct cluster recovery. These results differ considerably from those of other similarity studies. Blashfield

(1976a) found Ward's procedure to be the method of choice when clusters were generated in euclidean space using multivariate normal samples. Kuiper and Fisher (1975) came to the same conclusion.

More. Clustering techniques are usually sensitive to the presence of outliers or isolates that should be in no group. Also troublesome are intermediate entities that fit into several groups. Both Ward's minimum-variance method (Milligan, 1980) and NORMAP are badly affected by outliers (Everitt, 1974). To examine the effect of removing outliers prior to clustering, More (1981) evaluated the effectiveness of a three-stage clustering paradigm. The data sets were Fisher's iris data and a sample of 19 of Blashfield's computer-generated clusters. The data were first analyzed to obtain principal-component scores. Next Ward's minimum-variance procedure was applied after removing outliers. Finally a *k*-means algorithm in BMDP-71 was applied using the Ward findings as seed points. To detect outliers, Hawkins's (1974) procedure was followed. Three distances were computed and then used to test for significance. Outliers in factor-score space, in original variable space, and in residual factor space were removed. In the Blashfield clusters, removal of the multivariate outliers significantly improved the recovery accuracy of Ward's method and the *k*-means procedure.

Milligan. Milligan (1980) designed and conducted a study to examine the effect of error perturbation on 15 clustering algorithms. Artificial clusters that exhibited the properties of internal cohesion and external isolation were constructed. He used an error-free parent data set and five types of error perturbations (data sets with outliers, distance with chance error added, random noise dimensions, use of a noneuclidean distance measure, and standardization of variables). A three-factor design was followed: number of clusters, number of dimensions in which clusters were embedded, and three different patterns of point distribution. The clusters were generated in a euclidean space. But two noneuclidean distance measures were used on the error-free matrices: Cattell's r_p and the Pearson product-moment correlation converted by $D^2 = 2(1 - Q)$ to a dissimilarity measure. Rand's index was used as the external criterion of recovery; the internal criterion of goodness of fit was the point-biserial correlation.

All methods yielded mean recovery values of approximately 0.90 on the error-free condition except for MacQueen's algorithm. Complete linkage and Ward's method exhibited fairly noticeable decrements in cluster recovery under the two outlier conditions (20 percent and 40 percent); single linkage, group average, and the centroid method showed only slight decreases in recovery. The four nonhierarchical methods were virtually unaffected by the addition of outliers. Error perturbation of distances generated slight to moderate effects for hierarchical methods except for single linkage, which was strongly affected. The noneuclidean indices produced only slight decrements in recovery value.

In terms of rank order, group-average linkage placed among the top three methods in 8 out of 10 error conditions. The *k*-means method performed poorly in all error conditions. It is of interest to note that Ward's method, complete linkage, and single linkage did not even place among the top four procedures. The *k*-means procedure produced excellent recovery of structures when the starting seeds were obtained from the group-average method or when valid *a priori* information was available; moreover, the procedure was robust to all types of error. The internal criterion (point-biserial correlation) was successful in indicating the presence or absence of cluster structures in the data. It correlated 0.80 with the external criterion.

Summary of Monte Carlo Studies. Milligan's review of the earliest phase of the Monte Carlo studies (up to 1975) supported the view that Ward's minimum-variance-method gave the best recovery of cluster structures. The second phase (1975-1978) reflected an improvement in research design, data sampling, and range of algorithms compared. Again Ward's method, using euclidean distance, was superior in structure recovery. In the last phase (1979-1981) the five studies examined found that the group-average method was superior or equivalent to Ward's method.

The findings point to three critical factors: the similarity index, the extent of cluster overlap, and the treatment of outliers. Ward's method gives the best recovery in the presence of overlapping clusters. Group-average linkage provides superior recovery with nonoverlapping clusters that are well-separated; Ward's

method is most effective with a euclidean distance measure; average linkage provides superior recovery with correlations. Ward's method is most handicapped by clusters unequal in size and by the elevation dimension.

Constructed Data Set Studies

As was mentioned earlier, another approach to testing cluster structure is to contrive an artificial data set that mimics known classes. Following is a summary of two such studies.

Bartko and Colleagues. One of the earliest attempts to evaluate clustering techniques was applied to psychiatric data by Bartko, Strauss, and Carpenter (1971). Their data consisted of a group of 100 factitious archetypal patients. To obtain this group, ratings were fabricated to represent five diagnostic categories with 20 patients per category. The categories were neurotic depressed, paranoid schizophrenic, simple schizophrenic, manic depressive illness, manic type, and psychotic depressive reaction. A set of 251 items from the Present State Examination (PSE) was grouped into 48 units of analysis representing symptom patterns. In addition to varying the clustering techniques, the study assessed the influence of the type of input data used (dichotomous, polychotomous, or dimensional) and the merits of a factor-analytic data reduction.

Three cluster methods were compared: Friedman and Rubin's (1967) hill-climbing algorithm, Lorr and Radhakrishnan's average-linkage (nonhierarchical) procedure (1967), and McKeon's (1967) agglomerative complete-linkage (hierarchical) analysis. These three procedures were described in Chapters Five and Six. The hill-climbing method did not recover its original group, as three clusters were combined into one large cluster. When distances were used as measures of similarity in the hierarchical analysis, the results were not satisfactory. However, perfect replication of the input groups was obtained by using the monotonic transformation $2(1 - Q)$. The factor-score distance measure runs tended to distort the input groups for the hierarchical method. The average-linkage method was applied three times, varying the inclusion and exclusion cutoffs. The best output was with an 0.05 inclusion cutoff and 0.10 exclusion cutoff. All the depressed patients (neurotic and psychotic)

were put together, but the remainder were allocated correctly. The authors conclude that the best clustering was obtained by hierarchical complete linkage with correlation measures. In both methods, the factor scores drastically distorted the emerging clusters.

Mezzich. In a related study reported by Mezzich (1978), 10 clustering methods were evaluated: *Q*-factor analysis, multidimensional scaling, Chernoff's faces, single linkage, complete linkage, centroid average linkage, *k*-means, ISODATA, Rubin and Friedman's optimization techniques, and NORMAP/NORMIX. The coefficients compared were correlations, euclidean distances, and the city-block metric. Eight data sets based on random halves of four fields were used: psychosocial (72 treatment environments), psychopathological (88 archetypal psychiatric patients), botanical (150 iris specimens belonging to three species), and genetic (10 ethnic populations measured on 58 variables). The evaluative criteria were external validity (concordance of expected and actual), internal validity, and replicability. The complete-linkage method using correlation coefficients tended to rank highest. Next best were centroid average linkage using correlation coefficients and complete linkage using distances. The overall poorest result was obtained for NORMAP/NORMIX and single linkage.

Golden and Meehl. The purpose of the Golden and Meehl study (1980) was to evaluate the accuracy of common cluster methods when used to detect real empirical classes such as biological sex. The assumption was made that if a method fails to detect known taxa, it is unlikely to detect less understood taxa encountered in psychological and personality theories. Six cluster methods were subjected to trial: single linkage, complete linkage, average linkage, minimum variance, centroid linkage, and median linkage. Each cluster method was applied to four sets of sex-discriminant self-report items drawn from the MMPI. The validity of each item indicator was determined empirically, using biological sex as a criterion variable. The four sets of item indicators of biological sex were constructed so that they varied in their average item-indicator validity. Three cluster methods—complete linkage, average linkage, and minimum variance—detected accurately the biological sex taxonomy in most of the trials. Single linkage and the centroid and median methods were seldom accurate. The validity of the clusters

was assessed by means of the correct classification rate or *hit rate* (H) and by calculating coefficient kappa, which is equivalent here to $2H - 1$. The authors conclude (p. 475) that the cluster methods "are of questionable value since we typically lack assurance that the clusters are likely to be accurate and not spurious."

Summary

These evaluative studies make it clear that the problem of cluster recovery is indeed complex. Clusters may be spherical, ellipsoidal, or elongated, and they are embedded in an unknown number of attribute dimensions. The attribute dimensions or variables may vary in mean, dispersion, and degree of intercorrection; the clusters too may vary in degree of overlap and separation. Chance errors, outliers, and irrelevant variables all play a role in influencing cluster recovery.

Initially the Monte Carlo studies favored Ward's minimum-variance technique whereas complete linkage ranked second. The more recent studies, however, find that group-average linkage is most accurate among the hierarchical techniques. The minimum-variance procedure is too sensitive to the elevation parameter and is much affected by outliers. At the same time, as Milligan showed, all the hierarchical procedures are sensitive to outliers. For that reason some procedures are needed to remove outliers prior to clustering. The problem of elevation can be overcome by using a linear conversion of the Q correlations to distances. Although single linkage is repeatedly reported as least accurate in cluster recovery, it is not sensitive to outliers and has been shown to be useful for isolating elongated clusters or configurations such as circles.

The nonhierarchical clustering procedures of average, complete, and single linkage have not been subjected to much study. Nevertheless, the evidence is strong that k means and NORMAP/NORMIX should not be applied without a set of fairly stable seed points. Milligan (1980) has developed a two-stage procedure involving hierarchical average linkage followed by k means. Thus a useful procedure would be to remove outliers if possible, to apply, say, average linkage with a cutoff for the number of clusters, and then to compute a final version by use of k means.

The user should be constantly alert to the fact that clustering methods may impose structure on the data. The hierarchical agglomerative procedures, especially minimum variance, are biased toward circular or spherical clusters. Thus the investigator should try to use a procedure such as ordination to discover how the points are distributed in attribute space.

The next chapter is devoted to clarifying the role and value of *Q* analysis, which is often used to identify clusters. It shows that cluster analysis is designed to find discrete clusters while dimensional analysis isolates continua. Three kinds of *Q* or profile analysis are described and shown to be equivalent to corresponding *R* or variable analysis under certain conditions. Both result in dimensions, and the reciprocity principle for *Q* and *R* is explained and shown to be accounted for by the Eckart-Young theorem. Brief summaries of published *Q* analyses are presented to show the range of applications. The chapter ends with a step-by-step outline on how to conduct a *Q* analysis properly.

CHAPTER 8

Q Analysis

A *Q* analysis has often been proposed as a method for identifying mutually exclusive clusters or types. In this chapter we shall compare an *R* analysis of variables with a *Q* analysis of entity profiles. The discussion then turns to the uses of *Q* analysis. Its primary value in cluster analysis is to provide a low-dimensional mapping of a set of data points to facilitate cluster identification or confirmation. Another important use is for the analysis of *Q*-sort data and other types of rankings. If the purpose is to determine the dimensional structure of the attribute variables, *Q* analysis is shown to be appropriate when there are fewer entities than attributes. Under certain conditions, as we shall see, *R* analysis and *Q* analysis lead to comparable results. We shall also discuss the importance of following similar metricizing operations on the data matrix prior to *R* or *Q* analysis. The chapter concludes with a summary of studies illustrating common applications of *Q* analysis to data.

Cluster Analysis and Dimensional Analysis

Before contrasting *R* analysis and *Q* analysis it is useful to compare dimensional and typological analysis. As will be seen shortly, there is some dispute as to whether *Q* analysis is a method for

typing or a procedure for identifying dimensions. In what respects are cluster analysis and typological analysis different? As we have seen in prior chapters, cluster techniques are designed to find natural groupings or to create taxonomies. Dimensional analysis aims to isolate continua in terms of which all entities may be described. Most clusters are mutually exclusive; an entity either belongs to a cluster or it does not. The samples from populations studied are known to be, or suspected to be, multimodal. In a dimensional analysis, however, the multivariate distributions are ordinarily unimodal. If they were multimodal, interpretation of factors or components would be difficult and conflicting.

Another important distinction between factor-analytic and cluster-analytic techniques is in the treatment of the variance of a variable, whether it is an attribute or an entity profile. In factor analysis the variance is partitioned among several extracted factors or sources (Weiss, 1970). Factor analysis is called for when the goal is to simplify the attributes and express them in terms of a smaller number of common factors. In cluster analysis, on the other hand, the total variance of the entity is assigned to a discrete subgroup or cluster. Cluster analysis is most appropriate where discrete categories are sought. The cluster procedure reduces the complexity of a data set by sorting entities into a smaller number of homogeneous subgroups. The MMPI, for example, describes each subject in terms of 13 scores and results in a very large number of score profiles. As we saw in a prior chapter, most of these profiles can be classified into a half dozen homogeneous subsets.

R Analysis and *Q* Analysis

In order to characterize *Q* analysis it is necessary to distinguish it from *R* analysis. The distinction between the location of entities in attribute space and in entity space was discussed in Chapter Four, where we compared a dimensional analysis of the **R** matrix of attributes with a dimensional analysis of the **Q** matrix. To review this distinction recall the data matrix of N rows (entities) and n columns (attributes). The customary procedure is to correlate the attributes in the columns and then to factor-analyze the attribute intercorrelation matrix **R**. Such a factor analysis is called an *R*

analysis. If the rows (entities) are intercorrelated and the **Q** matrix of correlations is factored, the procedure is labeled a *Q analysis.* Naturally the mathematical theory behind these procedures is the same whether applied to the similarities between the columns or between the rows. Over time the analysis of entity profiles has come to be called *Q* analysis, *Q* technique, inverse analysis, and transpose analysis. Attributes are related over entities and entities are related over attributes. Hence one analysis is the transpose of the other.

Perhaps the first published factor analysis of correlations among persons was the one reported by Beebe-Center (1932). He conducted a study of the hedonic value of 14 olfactory substances determined by a panel of judges. Although Thomson (1950) says that he and Burt had conducted this sort of analysis for a number of years, the first popularization of the idea derives from Stephenson (1936). He argued that the method (called *Q* technique) was essentially different from a direct analysis of correlations among measures (called *R* technique). Stephenson contended that *Q* techniques yield types whereas *R* techniques identify dimensions. The *Q* correlations for 16 persons (manics, schizophrenics, and normals) over a sample of 30 moods were factor-analyzed to disclose three person factors.

The first dispute regarding the uniqueness of *Q* analysis was between Burt and Stephenson. Burt (1937) had developed a reciprocity principle in which he stated that the results would be equivalent if row and column means were subtracted from each score and each score were divided by the row and column dispersions. In his words (1940, p. 290), "the factor loadings for persons obtained by correlating persons are identical with the factor measurement for persons obtained by covariating tests." The principal difference is that factoring by tests retains a general test factor while losing a general entity factor. Factoring by entities retains a general entity factor but loses the general attribute factor.

The reciprocity principle was partially generalized to single centered and standardized matrices by Sandler (1949). Sandler showed that a principal-component analysis of sums of entity profile cross-products yields exactly the same results as a component analysis of product-moment correlations among attribute variables. Subsequently Harris (1953) provided a fully generalized solution. Ross (1963) has presented lucid illustrations of the equivalence of princi-

pal-component analysis by row and by column (providing the same data matrix is analyzed). He concludes that the nature of the dimensions does not change despite the different configurations formed by entities and attribute variables. Horst too discusses the problem of when to factor entities and when to analyze attributes (1965, pp. 324–326). He points out that the reciprocal relationships asserted by Burt and generalized by others hold only if the "basic structure" analysis (principal components) is performed without altering the diagonal elements of the product-moment matrix.

In a recent work Cattell (1978) asserts that *Q* technique is "not a method for finding types but for finding dimensions." Thus "the dimensions from *Q* technique are systematically related to and, with proper conditions, identical with those from the corresponding *R*-technique analysis" (p. 337). A fair number of methodologists do recommend *Q* analysis, however, among them Nunnally (1962), Guertin (1966), Holley (1970), Tryon and Bailey (1970), Overall and Klett (1972), and most recently Skinner (1979).

Types of Entity Profile Analysis

In a factor analysis or principal-component analysis it is the correlation coefficient that is ordinarily used to compare profiles or variables. But this choice may not be optimal because the correlation coefficient does not take scatter or level into account. *Covariances,* which are sums of deviation score cross-products divided by their number, can be used in this sort of analysis if only shape and scatter are of interest. Or one can factor-analyze raw score sums of cross-products that discard no information. As we saw in Chapter Three, distance is usually recommended as the index of similarity/difference. Because *D*'s are awkward to handle, however, sums of raw score cross-products (SSCP) provide an alternative (Nunnally, 1967). Analyses of *D*'s and SSCP's produce roughly equivalent results. Thus the researcher can apply principal-component analysis or factor analysis to correlation coefficients, covariances, or sums of raw score cross-products.

Which analyses are equivalent for entities and variables? It can be shown that corresponding to each analysis of entity profiles there is an equivalent principal-component analysis of variables.

Consider a data matrix **X** of N entities (rows) described in terms of n attributes (columns). The cross-product of **X** by its transpose **X′** yields the $N \times N$ cross-product matrix **XX′** showing to what degree the entities are related. A dimensional analysis of **XX′** represents a Q analysis. The cross-product of **X′** by its transpose **X** yields an $n \times n$ cross-product matrix **X′X** indicating how the attributes relate to each other. A dimensional analysis of **X′X** constitutes an R analysis. Usually a data matrix represents more entities (rows) than attributes (columns). It follows that the order (number of rows and columns) of **XX′** will ordinarily be greater than the order of **X′X**. But both matrices are square and symmetric. (For a further discussion of simple operations in matrices see Appendix B.)

It is useful to bring in the concept of *rank* at this point. Earlier we defined rank as the maximum number of linearly independent rows (or columns) of a matrix. More specifically the rank of the product of two matrices cannot be greater than the rank of the matrix of lowest order. Thus the rank of **XX′** is not ordinarily greater than the rank of **X′X** if the number of entities is greater than the number of attributes.

Let us now examine three types of Q analysis:

1. *Entity sums of score cross-products.* Suppose that the columns of **X** have been standardized to a mean of zero and a standard deviation of 1. The cross-product matrix of variable correlations is then

$$\mathbf{R} = \frac{1}{N}\,\mathbf{Z}'\mathbf{Z} = \frac{\mathbf{Z}'\mathbf{Z}}{N^{\frac{1}{2}}N^{\frac{1}{2}}}$$

where N, the number of entities, is a scalar (a real number). To simplify the right-hand term, multiply each element of **Z** by the constant $1/N^{\frac{1}{2}}$. Then

$$\mathbf{R} = \mathbf{Z}'\mathbf{Z}$$

and the standard deviation of each variable is $1/N^{\frac{1}{2}}$ instead of 1. If this standardized data matrix **Z** is postmultiplied by its transpose, we get the $N \times N$ matrix

$$\mathbf{CP} = \mathbf{ZZ}'$$

which consists of sums of raw score profile cross-products. A principal-component analysis of **CP** with sums of squares in the diagonal yields the same number of nonzero latent roots (eigenvalues) as for **R**. The principal components of **CP** correspond to the principal-component scores of the entities in **R**. As noted earlier, the rank of the product of two matrices cannot be greater than the rank of the factor of the lowest order.

2. *Entity covariances.* Again assume that the variables have been standardized to have means of zero and standard deviation of $1/N^{\frac{1}{2}}$. Let the entity profile means be set at zero by subtracting the entity means from each attribute scale value, but leave the row standard deviations unmodified. Here the cross-product matrix $\mathbf{Z'Z}$ for variables is not a correlation matrix because the profiles were mean-corrected. Now the cross-product of the mean-corrected entity profiles becomes

$$\mathbf{CV} = \mathbf{ZZ'}$$

which is the matrix of entity profile covariances. A principal-component analysis of **CV** yields entity components that correspond to the principal-factor scores of $\mathbf{Z'Z}$. The nonzero latent roots of **CV** are the same as the latent roots of $\mathbf{Z'Z}$.

3. *Entity correlations.* Assume, as before, that the variables have been standardized to a mean of zero and a standard deviation of $1/N^{\frac{1}{2}}$. Let the entity profiles also be centered and standardized to have means of zero and deviations of 1. Then multiply the elements of the matrix by the scalar $1/n^{\frac{1}{2}}$. Because the data matrix was altered by profile standardization, $\mathbf{Z'Z}$ is no longer a correlation matrix for variables. Now the cross-product of entity profiles yields

$$\mathbf{Q} = \mathbf{ZZ'}$$

which is the double-centered matrix first used by Burt to establish the reciprocity principle. As in the two previous cases, a principal-component analysis of $\mathbf{ZZ'}$ yields entity component coefficients that correspond to the attribute principal-component scores of the entities.

Thus one condition for equivalence is that the *same* data matrix, however metricized, forms the basis for the cross-product matrix analysis. If the data matrix used in a variable analysis is scaled differently from the data matrix basic to an entity analysis, the principal components and the components scores will be different. But for any given set of scaling operations it does not matter whether we factor variables or factor entities. For a given matrix we can solve for the component score matrix. Note, however, that the reciprocal relationships hold only if the principal-component (or basic structure) analysis is made without altering the diagonal elements in either case. If estimates of the communalities are entered in the cross-product matrices, this strict equivalence does not hold. Although the reciprocity principle holds for the score matrix of the common parts of the variables—that is, the matrix that generates the cross-product matrix with communalities—such a score matrix is not unique. There is only a rough similarity between **Q** and **R** factors if communality estimates are involved.

The reciprocity principle and the equivalence relationships between *R* analysis and *Q* analysis can be shown to derive from the Eckart and Young theorem (1936). Any matrix, such as $\mathbf{Z}$ $(N \times n)$, may be decomposed into a triple product called the basic structure matrix (Horst, 1963):

$$\mathbf{Z} = \mathbf{P}\boldsymbol{\Lambda}\mathbf{Q}'$$

where $\mathbf{P}$ $(N \times r)$ and $\mathbf{Q}$ $(n \times r)$ are orthonormal matrices and $\boldsymbol{\Lambda}$ $(r \times r)$ is a diagonal matrix with r latent roots.

The reader interested in the algebra of basic structure can find a clear exposition in Horst (1963). Simple illustrations of the process of finding eigenvectors or eigenvalues may be found in Green and Carroll (1976). Ryder (1964) presents an empirical example in which a data matrix for six subjects described on eight tests is analyzed in terms of both **R** and **Q**. Ross (1963) presents analyses of a contrived set of data for three entities described by four measures. Horst (1965) presents a worked-out example to illustrate the equivalence of direct and transpose basic structure analyses of a data matrix for 12 subjects described on nine tests. A simplified description of the Eckart–Young theorem expressed in matrix algebra is presented in a short appendix to this chapter.

An empirical illustration can be given that indicates how the dimensions isolated are similar whether identified by *Q* analysis or *R* analysis. Three subgroups of five individuals were selected from a sample of 90 to represent each of three body types hypothesized by Sheldon, Stevens, and Tucker (1941): endomorphs (fat), mesomorphs (muscular), and ectomorphs (slender) (Lorr and Fields, 1954). Each individual was assessed in terms of 36 morphological measures including height, limb length, weight, obesity, and muscularity of various limbs. An *R* analysis revealed three components. Table 8-1 presents the correlation of a selected set of 18 anthropometric measures with the morphological factors: Factor I is weight and bulk; factor II represents muscularity; factor III is defined by height and limb length.

For the *Q* analysis the 15 standardized score profiles were cross-multiplied. A *Q* analysis disclosed the three body types shown in Table 8-2. The first bipolar factor contrasts the obese endomorphs with the ectomorphs to define the weight dimension. Factor II contrasts the muscular mesomorphs and the frail ectomorphs.

Table 8-1. Correlation of Anthropometric Measures with Three Morphological Factors.

		Factors	
Measure	*I*	*II*	*III*
Weight	0.95	0.10	0.22
Transverse chest	0.91	0.13	0.05
Hip width	0.84	0.38	0.25
Sagittal chest diameter	0.94	0.12	0.07
Thigh diameter	0.97	−0.05	0.06
Height/weight	−0.94	0.02	0.22
Neck diameter	0.79	−0.31	0
Shoulder width	−0.02	−0.79	0.21
Muscle—trunk	−0.28	−0.83	−0.37
Muscle—legs	−0.08	−0.80	−0.11
Muscle—arms	−0.21	−0.89	−0.33
Height	−0.02	0.16	0.95
Trunk length	0.56	0.39	0.60
Arm length	−0.07	0.41	0.89
Symphysis height	−0.28	0.21	0.88
Sternal height	0.43	0.27	0.75
Leg length	−0.01	0.27	0.81
Foot length	−0.08	−0.17	0.88

Table 8-2. Body Type Factors.

Person	Factors I	II	III
Meso			
1	0.32	0.90	−0.23
2	0.10	0.68	0.18
3	0.34	0.75	0.03
4	0.23	0.25	0.52
5	−0.08	0.21	0.48
Endo			
6	−0.98	−0.10	−0.50
7	−0.64	0.17	−0.16
8	−1.14	−0.07	−0.08
9	−0.99	−0.37	0.13
10	−0.75	−0.37	0.11
Ecto			
11	0.87	−0.39	−0.46
12	0.79	−0.21	−0.27
13	0.76	−0.16	−0.31
14	0.65	−0.51	0.20
15	0.53	−0.76	0.38

Although factor III is less obvious (no photographs are given), the positively loaded individuals are tall and possess long arms and legs. Those who load negatively tend to be shorter than others in the sample. Thus factor III represents height/length. Both approaches, *R* analysis and *Q* analysis, clearly reveal similar morphological factors.

Illustrations of *Q* Analysis

This section presents brief summaries of *Q* analyses of a variety of research data involving normative measures. We shall also examine several other studies that exemplify the use of *Q* analysis for ranked data and measures collected by means of *Q* sorts.

A good example of a *Q* analysis may be found in French's (1970) study of motivational types among college students. The subjects were 86 students who completed 13 forms that generated 180 measures. The variables were first normalized to achieve a common metric. Entity relationships were assessed in terms of sums

of cross-products of the subjects across the 180 measures. The sums were then divided by an arbitrary constant to reduce them to a convenient size. A factor analysis of the matrix of sums of cross-products yielded 10 factors that were rotated obliquely by maxplane. The mean scores of the eight subjects who obtained the highest loadings and the eight who obtained the lowest were then computed for each factor as an aid in interpretation. Some of the isolated factors were labeled Moral Intellectuals, Hippie versus Conventional, and Emotionally Disturbed.

Tucker (1961) has reported an ingenious use of *Q* analysis to identify expert judges' basic value patterns. A panel of 19 experts was asked to evaluate the relevance of 225 items for assessing the ability of students in social studies. The items were rated on five-point scales. The mean cross-products of judges' ratings across the 225 items were factored with communality estimates in the diagonal cells of the matrix. The first factor was interpreted as a measure of general approval of the items. The second factor indicated sharp differences of opinion between two groups of judges. One group valued information items; the other valued ability to organize material and reach generalized conclusions.

Jay's (1969) *Q* analysis of the Rokeach dogmatism scale represents another often-quoted report. The 40-item scale was administered to 29 students. Each item was rated on a six-point scale of agreement/disagreement. The students were intercorrelated across the 40 items, and a principal-component analysis was made of the matrix. The three factors were interpreted in terms of the items on which the three persons with the highest loadings agreed. The factors were interpreted as open-mindedness, true believers motivated by fear, and belief in a single cause. Jay interpreted the factors as types.

In early *Q* analysis conducted by Coombs and Satter (1949), the data were collected in the course of evaluating 70 jobs. Employees were interviewed to prepare job descriptions that were then recorded on a rating sheet calling for judgments of various skills and information. Included were 104 educational skills, work skills, application skills, social and personnel skills, and activities. To reduce the labor of analyzing 70 occupations a submatrix of order 20 × 20 was selected. The correlations among the 20 occupations were based on

a common-element correlation formula. The matrix was factored by the centroid method to identify six factors: self-responsible jobs (librarian), routine entry jobs (truck driver), skilled machine operation (multigraph operator), clerical jobs (record clerk), and a general factor. The large number of job elements justified the use of *Q* correlations. The appearance of a general factor is probably due to failure to convert the data into deviation form.

Overall and Klett (1972) endorse *Q* analysis as a method for defining ideal types. A good example is provided in one application to psychiatric patients. The data analyzed were ratings by a group of 38 American experts who evaluated 13 diagnostic classes on the 16 scales of the Brief Psychiatric Rating Scale. *Q*-type correlations were computed among the 13 diagnostic profiles across the 38 judges. A principal-component analysis of the correlation matrix disclosed three superordinate classes: paranoids, schizophrenics, and depressives. Overall and Klett conceptualize *Q* analysis as "linear typal analysis" that seeks to discover the nature and number of underlying pure types. Within the context of the linear factor model they think it reasonable to conceive of person factors as ideal types and to regard factor loadings as indices of relationship of individuals to several ideal types. The notion of ideal types has also been adopted by Skinner (1978). However, an *R* analysis of the same data matrix is highly likely to lead to comparable dimensional constructs.

Skinner and Jackson (1978) investigated the degree to which MMPI profiles can be conceptualized in terms of three superordinate types (neurotic, psychotic, and sociopathic). A second purpose was to evaluate the Gilberstadt and Duker (G/D; 1965) and the Marks, Seeman, and Haller (M/S/H; 1974) code-type system in terms of these superordinate types. The three prototypes were conceptualized as ideal type constructs. The data analyzed were the 35 MMPI code-type profiles, 19 from G/D and 16 from M/S/H. A principal-component procedure was applied to the correlations among the 35 code-type profiles. The three modal profiles they derived were interpreted as neurotic, psychotic, and sociopathic ideal types. The authors emphasize the need to differentiate profile elevation, scatter, and shape in matching a given MMPI profile to the various code types. They explain that while relatively homogeneous subgroups or clusters can be identified in the ordination space,

their approach is essentially a dimensional model. A profile was allocated to that entity factor on which it loaded highest and above 0.50 regardless of algebraic sign. In terms of their criteria, 84.6 percent of the 233 Lanyon (1968) groups were classified as salient on a specific modal profile.

Q-Sort Analysis

An important use of *Q* analysis is to examine *Q*-sort data and other types of ranked data such as pair comparisons. In the *Q-sort method,* the rater is given a set of statements or items previously structured to represent some universe of attributes. The rater is then asked to characterize a particular entity (object, person, stimulus) in terms of the set of items. The items are ranked, from most to least with respect to representation, salience, or significance for the entity. The rater must order the items into a designated number of categories, usually 9 or 11, with a specified number of items placed in each category. The distribution is typically quasi-normal—for example, there might be 5, 8, 12, 16, 18, 16, 12, 8, 5 items in a sequence of nine categories. Items most characteristic of the person are placed at one end of the continuum; items most uncharacteristic are placed at the other end. In a *Q* analysis, members of a sample of entities are intercorrelated across the item set. The correlation matrix is then factored to identify any types that are present.

Note that *Q* analysis rather than *R* analysis is more appropriate to *Q*-sort data because the items are ranked. Each score for a person is dependent on his or her scores on other variables but is independent of the scores of other individuals. Such scores, called *ipsative,* may be contrasted with the properties of *absolute* measures that are statistically independent of the individual's other scores. There are two types of absolute measures: *raw* and *normative.* Since a raw score cannot be interpreted as high or low, scores are compared to the average performance of members of a sample that constitutes the norm. Here an individual's position is statistically dependent on the scores of other individuals in the sample. In any case scores that are obtained as ipsative measures may legitimately be employed only for purposes of intraindividual comparisons. By stringent definition a score matrix is ipsative if the sum of the scores obtained over the

attributes for each person is a constant. Each person has the same mean since the attribute scores sum to the same constant. For *R* analysis, normative scores are appropriate because scores are correlated over a population of individuals. For *Q* analysis involving correlations between individuals, ipsative scores, or scores converted to ipsative form (normative-ipsative), are necessary. Then two or more individuals with similar intraindividual score patterns correlate highly and are grouped together.

Stein and Neulinger (1968) have developed a typology of self-descriptions based on Stein's Self-Description Questionnaire (SDQ). The SDQ consists of 20 paragraphs describing the 20 manifest needs defined by Murray (achievement, aggression, dominance, order, and so forth); the subject is asked to rank the paragraphs from most descriptive to least descriptive. The SDQ was administered to 80 Peace Corps volunteers, 116 chemists, and 115 male students at NYU. The product-moment correlations of subjects within each sample were analyzed by principal components and a *varimax rotation* was applied. The *Q* analysis yielded four, five, and seven factors or types within the three samples. The profile of a type was determined by the mean ranking of needs by the *type definers*—that is, the subjects who loaded highest on a given factor. The 16 type profiles were then matched and reduced to five: socially oriented, intellectually oriented, action oriented, unconventional, and resourceful. Since for every *R* type there is a corresponding *Q* analysis, it would have been possible to infer the groupings of manifest needs measured by the 20 paragraphs.

The use of *Q* analysis to study the manifest needs of groups rather than individuals has been reported by Gordon and Sait (1969). The Edwards Personality Preference Schedule, a forced-choice ipsative form, was administered to 45 groups of males and females: prisoners, hospitalized psychiatric patients, military personnel, industrial employees, and students and teachers. The groups were correlated on the basis of their mean scores on the 15 need scales. The correlation matrix was analyzed by the principal-component method. Four factors were retained and interpreted in terms of group factor loadings and the scale means of these groups best defining each factor. The factors were labeled docility, dominance striving, friendly interest, and adolescent revolt.

Conducting a *Q* Analysis

Q analysis is a frequently applied procedure. It is useful for ordination, for the analysis of ipsative data, and for dimensional analysis if the entity sample is smaller than the variable sample. Further, it is evident from the literature that *Q* analysis is used to isolate types. Accordingly the procedure will be outlined here in some detail—the following sections discuss the five basic steps. Note that the term *Q* analysis covers not only the analysis of entity intercorrelations but also covariances and sums of raw cross-products.

Standardizing Variables. Usually it is best to center and standardize variables across entities before calculating *Q* correlations or sums of raw cross-products. Different elements (variables) within the profile may have substantially different means and variances across entities. In a morphological study, for example, the profile elements may differ in units such as weight in pounds and height in inches. Or the dispersions may differ for, say, length of arm and width of head. A uniform metric can be established by subtracting the column (variable) means from all the columns and then dividing all the corrected column values by their dispersions. In this way all the variables will uniformly have zero means and unit variances. Standardization of variables is equally important when the data consist of derived standard scores as found in the MMPI. The norm sample is given a mean of 50 and a standard deviation of 10. However, the sample scales analyzed may have means and dispersions that differ widely and thus require restandardization. When *Q* sorts like those developed by Stephenson (1953) are used, the items are sorted into quasi-normal or rectangular distributions. Although the variables are not usually standardized, the scaling procedure should be applied if the variables are badly skewed.

Assessing Similarity. Let us assume that the variables in the data matrix have been standardized to a mean of zero and a sigma of 1. As explained earlier in the chapter, the similarities between entities can then be computed in several ways. One procedure, advocated by Nunnally, Tucker, and Ross, is to compute the sums of cross-products and sums of squares of the profile measures. If a strong general factor such as acquiescence or social desirability is present, however, this approach may not be best. The usual procedure is to correlate

the rows (entities). In the process of correlation the means of the row are subtracted from each of the n variables that define the profiles. The resulting deviation scores within each row are then multiplied by the reciprocals of the variable standard deviations. The between-entity correlation is then $\mathbf{Q} = (1/n)\mathbf{ZZ}'$. The result is a matrix approximately double-centered.

Factoring the Similarity Matrix. Once the matrix of cross-products or the matrix of Q correlations is available, a choice must be made regarding the method of dimensional analysis. Principal-component analysis, the most common procedure, is particularly appropriate if one is interested in a solution equivalent to an R analysis. Moreover, in practice principal-component analysis yields factors quite similar to those identified by principal axes, the method that calls for communality estimates. A further reason is that principal-component scores, if needed, can be computed readily.

An associated problem concerns the use of criteria for deciding when to stop factoring. A fairly sound and frequently used criterion is known as the *scree test* (Cattell, 1966). In this procedure the eigenvalues (latent roots) are plotted with the value of the root along the ordinate and the factor ordinal number as the abscissa. A straight line is fitted to the bottom portion of the curve. The point where there is an inflection in the curve gives the number of components to retain. The number of eigenvalues greater than 1 is another criterion, but it is a doubtful guide. An extended discussion may be found in Harman (1976) and in Gorsuch (1974).

Finding a Rotational Solution. In order to interpret the entity factors, it is necessary to find a unique solution or simple structure. Generally Kaiser's *normal varimax* is applied to obtain an orthogonal interpretive framework. If the orthogonal solution offers a poor fit as judged by the number of complex entities (loadings on more than one factor), then a correlated or oblique solution should be sought. A quick and widely accepted method called *promax* (Hendrickson and White, 1964) can be applied. Choice of the rotational solution is an important consideration because the procedure may substantially affect factor loadings, which provide the basis for allocating entities to clusters.

Allocating Entities to Clusters. To assign entities to clusters (subgroups), criteria are needed to specify who belongs. The most common procedure is to assign entities to a group on the basis of a

Table 8-3. Correlations Among 15 Subjects.

Person	*1*	*2*	*3*	*4*	*5*	*6*	*7*	*8*	*9*	*10*	*11*	*12*	*13*	*14*	*15*
Meso															
1		0.73	0.79	0.34	0.18	−0.21	−0.01	−0.27	−0.49	−0.59	−0.22	−0.07	0.06	−0.39	−0.54
2			0.75	0.36	0.51	−0.29	0.02	−0.20	−0.42	−0.51	−0.34	−0.27	−0.10	−0.33	−0.47
3				0.65	0.29	−0.42	−0.14	−0.36	−0.60	−0.70	−0.19	0.03	0.07	−0.21	−0.41
4					0.25	−0.73	−0.35	−0.59	−0.65	−0.60	0.08	0.27	0.31	0.28	0.14
5						−0.19	−0.13	−0.17	−0.14	−0.18	−0.43	−0.17	−0.19	−0.17	−0.01
Endo															
6							0.64	0.85	0.86	0.72	−0.53	−0.66	−0.72	−0.69	−0.52
7								0.70	0.49	0.55	−0.63	−0.69	−0.63	−0.71	−0.64
8									0.80	0.76	−0.58	−0.76	−0.77	−0.65	−0.57
9										0.83	−0.46	−0.64	−0.72	−0.43	−0.25
10											−0.39	−0.56	−0.62	−0.29	−0.16
Ecto															
11												0.79	0.76	0.79	0.64
12													0.80	0.73	0.72
13														0.68	0.61
14															0.83
15															

Table 8-4. Correlation of Persons with Orthogonal and Rotated Oblique Person Factors.

	Orthogonal Factors			Oblique Factors	
Person	*I*	*II*	*III*	*A*	*B*
Meso					
1	0.34	−0.65	−0.44	0.01	0.62
2	0.33	−0.68	−0.44	−0.02	0.64
3	0.50	−0.62	−0.44	0.17	0.64
4	0.66	−0.24	−0.18	0.48	0.38
5	0.16	−0.38	−0.26	−0.02	0.35
Endo					
6	−0.93	−0.08	−0.06	−0.87	−0.20
7	−0.74	−0.28	−0.22	−0.78	0.02
8	−0.93	−0.12	−0.09	−0.88	−0.16
9	−0.91	0.09	0.09	−0.77	−0.33
10	−0.87	0.15	0.16	−0.71	−0.37
Ecto					
11	0.55	0.42	0.32	0.68	−0.18
12	0.73	0.32	0.26	0.79	−0.05
13	0.77	0.25	0.20	0.79	0.02
14	0.61	0.44	0.37	0.75	−0.19
15	0.47	0.50	0.30	0.64	−0.27

correlation of |0.50| or |0.60| (absolute value). An additional and equally important criterion is that the entity must not load on any other factor above a given maximum value (say ± 0.35). To establish mutually exclusive groups, an entity should not be allocated to more than one group. The homogeneity of the clusters as well as their coverage is influenced by the cutoff criteria. A low cutoff point yields a classification of a high proportion of entities but a heterogeneous subgroup; a high cutoff point results in low coverage but a highly homogeneous subgroup. For these reasons the investigator should compute a measure of internal consistency among members of a cluster to judge the appropriateness of the cutoff value.

The bipolar nature of entity factors poses a related problem. Setting entity means to zero and then factoring variables shifts the origin to the centroid of the *variables.* But setting variable means to zero and then factoring entities shifts the origin to the centroid of the *entities.* Since *Q* analysis is usually carried out with double centering, most *Q* factors are bipolar. This is why the cutoffs disregard the algebraic sign of the factor loading. Overall and Klett (1972) suggest that a constant be added to all mean-corrected scores prior to factor analysis. The ideal origin will yield a $\mathbf{Q} = \mathbf{ZZ}'$ matrix on which elements are predominantly positive (p. 222).

Earlier in the chapter we examined a principal-component analysis of sums of cross-products of five mesomorphs, five endomorphs, and five ectomorphs. Now we shall consider the results of a factor analysis of the double-centered matrix of morphological measures. Table 8-3 presents the *Q* correlation among the 15 subjects. It can be seen that intercorrelations between the body types tend to be negative—as one can expect when data are double-centered. Table 8-4 presents the unrotated and the rotated table of factor loadings on the person factors. It can be seen that rotation does not remove bipolarity. Factor I contrasts endomorphs and ectomorphs whereas factor II isolates the mesomorphs. Thus from the viewpoint of the typologist factor I identifies two subtypes.

Summary

A *Q* analysis has often been proposed as a method for identifying clusters or types but the appropriateness of the method for this purpose has long been in dispute. To clarify the issue, cluster analysis

and dimensional analysis are compared. Cluster techniques are designed to find discrete groupings or to create taxonomies. Dimensional analysis isolates continua called factors, in terms of which all entities can be described. The origin of *Q* analysis is traced and its relation to *R* analysis is made explicit. The reciprocity principle developed by Burt and generalized by Harris and others, is then linked to the Eckart and Young theorem and the basic structure of a matrix. Both *R* and *Q* analysis are interpreted as primarily methods for finding dimensions, not for finding types. Three kinds of entity profile analysis (*Q* analyses) are described. One is based on correlations, another on covariance, and a third on sums of cross-products. It is shown that corresponding to each analysis of entity profiles there is an equivalent principal components analysis of variables, providing the diagonal elements of the matrix of similarities are left unaltered. Several concrete illustrations show how the dimensions isolated are similar whether identified by *Q* or *R* analysis, summaries of published *Q* analyses are presented, and a step-by-step explanation of how to conduct a *Q* analysis is given.

Appendix: Basic Structure of a Matrix

The reciprocity principle and the equivalence relationship between analyses of variables and entities derive from the Eckart and Young theorem (1936). According to Horst (1963), any nonhorizontal matrix $\mathbf{Z}$ $(N \times n)$ can always be expressed as the product of a nonhorizontal matrix $\mathbf{P}$ by a diagonal matrix $\mathbf{\Lambda}$ by the transpose of the nonvertical orthonormal matrix $\mathbf{Q}$. In other words any matrix may be decomposed into a triple product

$$\mathbf{Z} = \mathbf{P\Lambda Q'}$$

Here $\mathbf{P}$ $(N \times r)$ is called the left orthonormal of $\mathbf{Z}$; $\mathbf{\Lambda}(r \times r)$ is the basic diagonal of $\mathbf{Z}$; and $\mathbf{Q'}(r \times n)$ is the right orthonormal of $\mathbf{Z}$. Since $\mathbf{P}$ and $\mathbf{Q}$ are orthonormal, $\mathbf{P'P} = \mathbf{I}$ and $\mathbf{Q'Q} = \mathbf{I}$, where $\mathbf{I}$ represents an identity matrix. In an orthogonal matrix each column vector is orthogonal to every other column vector. An orthonormal is an orthogonal matrix in which all column vectors are normal (of

unit length). The diagonal elements of $\mathbf{\Lambda}$ are nonzero and give the square roots of the r eigenvalues. Representation of matrix $\mathbf{Z}$ as a triple product is called its *basic structure* by Horst; numerical analysts sometimes call it a *singular value decomposition.* The theorem gives a least-squares estimate of the rank of $\mathbf{Z}$ (or of any matrix). By *rank* is meant the maximum number of linearly independent rows and columns of the matrix.

It can be shown that a principal-component analysis of a matrix of variable intercorrelations is equivalent to using Eckart and Young's model to decompose an $N \times n$ matrix $\mathbf{Z}$. The entries of $\mathbf{Z}$ are the product of $1/N^{\frac{1}{2}}$ times the standardized scores for N entities on n variables (assuming $N \geq n$). To find the basic structure, the product $\mathbf{Z'Z} = \mathbf{Q\Lambda^2Q'}$ is computed. A principal-component solution of $\mathbf{Z'Z}$ yields the eigenvalues Λ^2 and the matrix $\mathbf{Q}$ of associated eigenvectors. Then, knowing $\mathbf{Q}$, solve for

$$\mathbf{P} = \mathbf{Z Q \Lambda^{-1}}$$

If $\mathbf{P}$ is known, one can solve for $\mathbf{Q}$:

$$\mathbf{Q} = \mathbf{Z' P \Lambda^{-1}}$$

The triple product $\mathbf{Z} = \mathbf{P\Lambda Q'}$ may be expressed as $\mathbf{Z} = \mathbf{P\Lambda^{\frac{1}{2}}\Lambda^{\frac{1}{2}}Q'}$. Then the familiar factor matrix of factor coefficients is

$$\mathbf{A}_t = \mathbf{Q\Lambda^{\frac{1}{2}}}$$

and the factor-score matrix is

$$\mathbf{F}_t = \mathbf{P\Lambda^{\frac{1}{2}}}$$

These matrices are linked by the fundamental factor theorem (see Chapter Four) as follows:

$$\mathbf{Z'} = \mathbf{A}_t\mathbf{F}'_t$$

in which $\mathbf{Z}$ is $n \times N$, $\mathbf{A}_t$ is $n \times r$, and $\mathbf{F}'_t$ is $r \times N$.

The equivalence relations between R analysis and Q analysis

can be summarized by recalling the procedure:

- A principal-component analysis of $\mathbf{R} = \mathbf{Z}'\mathbf{Z}$ yields a test (variable) factor-loading matrix $\mathbf{A}_t$ from which a factor-score matrix $\mathbf{F}_t$ for entities can be derived.
- A principal-component analysis of $\mathbf{Q} = \mathbf{Z}\mathbf{Z}'$ results in a profile factor-loading matrix $\mathbf{A}_p$, from which a matrix of entity factor scores $\mathbf{F}_p$ for tests (variables) may be computed.

The test (variable) factor-*loading* matrix $\mathbf{A}_t$ is equivalent to the profile factor-*score* matrix $\mathbf{F}_p$. Similarly the factor-loading matrix $\mathbf{A}_p$ for entities is equivalent to the test (variable) factor-score matrix $\mathbf{F}_t$. Another solution is to compute $\mathbf{A}_t$ and solve for its equivalent matrix $\mathbf{A}_p = \mathbf{Z}\mathbf{A}_t\Lambda^{-\frac{1}{2}}$ or to get the equivalent of $\mathbf{A}_p$ and solve for $\mathbf{A}_t = \mathbf{Z}'\mathbf{A}_p\Lambda^{-\frac{1}{2}}$.

CHAPTER 9

Using Subgroups for Prediction

Chapter One described the goals of typological analysis. The first four objectives emphasized the discovery or construction of homogeneous groupings. The fifth goal was to identify homogeneous subgroups in which members are similar in their predictor profiles. Once identified these subgroups can be used to predict the future behavior of subgroup members. A variant of this conception is known as the actuarial model for prediction. The advantages claimed for the model are twofold: greater predictive efficiency because of possible configural relations and greater understanding of the manner in which the predictor variables interact. This chapter outlines the various steps of the actuarial approach and presents illustrative studies in several areas. In addition we shall examine the comparative effectiveness of traditional multiple regression and prediction via subgroups.

The Actuarial Model

It is generally recognized that there are several approaches to prediction. The conventional approach is to apply multiple regression to a set of predictors of a dependent measure. This traditional model has at times been judged inadequate because of its failure to

take into account higher-order terms and interactions. A second approach involves grouping entities into homogeneous classes likely to be useful for prediction. Toops (1948, 1959) helped develop this model. He believed that society could be grouped into homogeneous subsets of individuals or *ulstriths.* His argument was that such subgroups have similar biographies and experiences and will therefore exhibit similar patterns of future behavior. There are two ways to use taxonomic classes. One approach is to use the *actuarial model.* By this method probability estimates of future behavior or outcomes are derived from contingency tables that relate the taxonomic classes to the criterion variables. The second approach is to use multiple regression to predict the dependent variable from knowledge of membership in the various subgroups.

Linear multiple regression and its various forms (Cohen and Cohen, 1975) are the principal methods for weighting and combining continuous predictor variables in order to predict continuous dependent measures (called *criteria*). There are many cases, however, in which the predictor variables are categorical (marital status, religion, socioeconomic class). Moreover, the forecast concerns the probability that an individual with a certain combination of characteristics will actually achieve a certain criterion status—graduation from college, risk of an auto accident, presence or absence of given psychiatric diagnosis, selection or rejection for a training program, and so on. The procedures that determine such probability estimates of criterion status from contingency tables relating predictor and criterion variables are referred to as *actuarial prediction* (Sines, 1966; Wiggins, 1973).

The use of actuarial prediction is common in the insurance industry for life insurance, auto accident insurance, and other forms. Moreover, actuarial methods are the primary means for determining release of a prisoner on parole. In most of these cases data are collected on categories such as age, sex, type of community, educational level, number of prior accidents, number of arrests, and number of hospitalizations. The prediction problem may be stated by use of notation suggested by Dawes (1962). Stated formally the predictor variables may be represented by the symbols R_1, $R_2, \ldots, R_j, \ldots, R_k$, which could imply responses to a psychological test or membership in a dichotomous or polychotomous

variable. The external criteria (dependent variables) may be represented by the symbols $C_1, C_2, \ldots, C_i, \ldots, C_k$. The prediction problem then takes the following form: Given a series of predictors $R_1, R_2, \ldots, R_k$, what is the probability that an entity with this pattern of responses will be a member of a criterion group $C_1, C_2, \ldots, C_k$? The research problem is one of estimating the probability of criterion group membership from the observed frequencies in the contingency table. Once validity data have been collected, the inference is of the form $P(C_i|R_j)$—that is, given a particular pattern of characteristics or test responses, what is the probability that the individual with this pattern R_j is a member of a specific criterion group C_i?

Given that an individual is a member of a criterion group C_i, what is the probability that his or her characteristics will be R_j? Stated formally: $P(R_j|C_i)$. A comparison of the expression $P(R_j|C_i)$ with the form $P(C_i|R_j)$ suggests that the predictive inferences are quite different. As Dawes and others point out, it is by no means certain that

$$P(C_i|R_j) = P(R_j|C_i)$$

To the degree that the values are not equivalent the validity information given in a test manual is not useful for clinical prediction.

To establish an actuarial prediction system, it is vital to choose a set of variables that are relevant to the predictions to be made. Moreover, they should be objective, reliable, representative of the domain being studied, and independent of each other. Some method of cluster analysis is then chosen to generate the needed classification. Account must also be taken of the index of profile similarity. As discussed in earlier chapters, the issue is whether the elevation parameter or scatter is relevant or should be eliminated. Once the taxonomic subgroups have been established, it becomes possible to construct actuarial prediction tables from the criteria data. The problem thus turns to the criteria themselves.

One kind, called *fixed* criteria, is illustrated by success/failure in a training course, completion of college, release from the hospital, or suicidal risk. A second kind may be thought of as open-ended or *free* criteria (Gleser, 1963)—that is, predicting from a taxonomic class (Sines, 1966). In this case the focus is on score patterns that

enable the user to describe a person and in fact use the subgroup for a wide range of predictions. Sines and others suggest that reliance on a fixed criterion involves a prohibitively large amount of time and energy because each criterion requires identification and validation of a separate set of predictors.

If free criteria are used, the first task is to establish a taxonomic classification that will provide a manageable number of homogeneous and mutually exclusive classes occurring with sufficient frequency to provide a reliable basis for prediction. The second task is to determine the extent to which class members share similar etiologies, treatment response patterns, and so on. Actuarial prediction tables are constructed from the biographical data, demographic data, personality descriptors, diagnoses, and so on. The frequency of occurrence of different categories is tabulated separately for each class. These frequencies in turn may be converted to percentages.

It should be recognized, however, that a classification is useful only to the extent that it is based on variables related to the broad class of behavior one is trying to predict or control (Gleser, 1963). The taxonomy must be natural—that is, general purpose as defined in earlier chapters.

Actuarial Prediction Systems

Several actuarial prediction systems using the Minnesota Multiphasic Personality Inventory (MMPI) have been constructed. The two best-known systems were developed by Gilberstadt and Duker (1965) and by Marks and Seeman (1963). The MMPI profiles are based on 13 scale scores expressed in standard-score form with arbitrary means of 50 and standard deviations of 10. Instead of applying clustering techniques, Gilberstadt and Duker established homogeneous subgroups by locating prototypical examples of profile types with classic case histories. Then preliminary rules were established to specify who belonged to a code type. Next they scanned case history records to check on the accuracy of the rules. Finally they defined 19 profile types based on the mean profile of groups of case records judged to represent classic cases. Each profile type included a minimum of nine cases. The 10 clinical scales are numbered sequentially: 1, 2, 3, 4, 5, 6, 7, 8, 9, 0. Each profile type

was labeled in order of the scale score elevations, from highest to lowest, that exceeded a T score of 70. If scales 2 (depression), 7 (psychasthenia), and 8 (schizophrenia) are elevated beyond 70 in the order given, for example, the profile is called a 2-7-8. Another example is the profile type 1-2-3, where 1 (hypochondriasis), 2 (depression), and 3 (hysteria) are greater than 70. No other scales should be greater than 70; the requirements for the validity scales are $L \leq 65$, $R \leq 85$, $K \leq 70$.

Once the 19 profile code types were established, data were assembled on (1) the most probable psychiatric diagnosis and alternative diagnoses, (2) characteristic complaints, traits, and symptoms, (3) the cardinal clinical features of the profile type, and (4) descriptive material on background and early history, vocational and educational adjustment, and clinical appearance of the criterion group. In applying the system the researcher scores the MMPI profile and looks for a similar profile among the 19 illustrated in the handbook. When a similar profile is located, the specification rules are applied to see whether they are satisfied. Finally the researcher checks the characteristics associated with the subgroup.

Marks and Seeman examined the MMPI profiles of 248 patients at the University of Kansas Medical Center and tentatively identified nine profile types. The rules for these types were refined by application to 387 other psychiatric patients, and seven profiles were added. Later a sample of 826 was used to establish mean profiles for 16 types, each based on 20 or more cases. Descriptors were based on 108 *Q*-sort items. Another source of information consisted of ratings of hospital records for the presence or absence of 225 case history variables. Group differences were tested by chi square and *t*-tests. Demographic data (age, sex, marital status, number of children) were also collected.

The Marks/Seeman system can be applied either by hand scoring or by an automated computer procedure. Comparison of the profile specification rules of the two systems reveals that Marks and Seeman's requirements are much more narrowly defined than those of Gilberstadt and Duker. Either system alone classifies approximately 28 percent of a patient population whereas joint application results in a classification rate of 49 percent (Payne and Wiggins, 1968).

Linear and Configural Models

Multiple regression and correlation is a well-established data-analytic procedure in the social and behavioral sciences. The procedure is applied when the relationship between one variable (the criterion or dependent variable) and a set of two or more variables (independent or predictor variables) is to be examined. The method has recently been much expanded in scope and generality. By coding information, the procedure has been expanded to include nominal or qualitative scales and nonlinearly related quantitative variables and interactions.

The customary form of a linear multiple-regression equation is as follows:

$$Y' = B_1X_1 + B_2X_2 + B_3X_3 + \cdot \cdot \cdot + B_nX_n + A$$

where Y' is the predicted criterion variable, $X_1, X_2, \ldots, X_n$ are the independent variables, and $B_1, B_2, \ldots, B_n$ are the partial regression coefficients or weights. These estimates are the best possible by the least-squares criterion. This means that if the error in estimating an entity's status is measured by the discrepancy between actual status and status as estimated by the equation Y', the sum of these squared errors $(Y - Y')^2$ would be smaller than by any other set of constants. The multiple correlation (R) of a criterion with a set of n predictors ($X_1, X_2, \ldots, X_n$) is symbolized as $R_{y.12\ldots n}$—that is, the product-moment correlation between the actual criterion value Y and the estimated criterion value Y' obtained from the regression equation.

It appears that Meehl (1950) first introduced the *configural scoring* hypothesis. The idea is that useful information may be contained in the pattern of response to a pair of dichotomous items even when the responses taken singly contain little or no differential information. For example, answering T (true) to both items 1 and 2 on a test (T_1T_2) or F (false) to both items (F_1F_2) may identify members of group A. Answering T_1F_2 or F_1T_2 may identify members of group B. Taken additively, $T_1 + T_2$ might not differentiate at all. Horst (1954) suggested that the configural scoring problem should be conceptualized as determining the significance of the interaction

terms like (X_1X_2) in the multiple-regression equation

$$C' = b_0 + b_1X_1 + b_2X_2 + b_{12}X_1X_2$$

where X_1 is the score on item 1, X_2 is the score on item 2, C' is the predicted criterion score, and the b's are regression coefficients.

Lubin and Osburn (1957) developed a similar theory of pattern analysis for the prediction of a quantitative criterion from dichotomous items. They define a configural scale as a set of 2^t criterion averages, one for each answer pattern. The scale can be represented as a polynomial function (one with several terms) of the item scores as follows:

$$\begin{aligned} C' = b_0 + b_1X_1 + b_2X_2 + \cdots + b_tX_t \\ + b_{12}X_1X_2 + \cdots + b_{123}X_1X_2X_3 + \cdots \end{aligned}$$

The configural scale was shown to have maximum validity in the least-squares sense. The cross-products represent possible interaction terms.

The configural scoring procedure is not restricted to dichotomous items. It can be extended to polychotomous items, but there are several difficulties with this approach. The principal disadvantage is that the number of answer patterns is very large. Even when the number of dichotomous items is as small as 10, for example, 2^{10} is 1,024. The procedure is most appropriate for situations involving few variables but many entities. One method often employed is to use only a specially selected subset in an analysis. Another important disadvantage is that significant nonlinear or cross-product terms are difficult to cross-validate. Seldom can an investigator find the same significant cross-product terms in a second sample.

The Goldberg Studies

Insight into how configural relations among predictors contribute to prediction may be gained from a series of studies by Goldberg (1965, 1969). At an Educational Testing Service conference in 1965 Meehl observed that the relationship between MMPI scores and psychosis versus neurosis diagnostic classification should

be highly configural. Therefore he predicted that no linear combination of MMPI scores would be able to differentiate neurotic from psychotic patients as accurately as configural actuarial techniques or experienced psychologists. Challenged by this assertion, Goldberg (1965) reexamined the 861 MMPI profiles obtained by Meehl from seven hospitals and clinics, as well as the diagnostic judgments of 29 clinical judges. The accuracy of each of these psychologists was compared with the validity achieved by each of 11 MMPI scale scores, 8 scale ranks, 54 diagnostic signs and rules, 35 profiles, 19 linear regression analyses, and numerous actuarial tables.

The cross-validity of each indicator and each judge was ascertained for each of the seven samples. Goldberg found that a simple linear combination of five single scales from the MMPI outperformed all 29 diagnosticians. Since the initial publication of these findings Goldberg (1969) and his associates have conducted seven additional studies. They found that neither a multiscale moderator variable nor the profile code types developed by Marks and Seeman (1963) and by Gilberstadt and Duker (1965) exceeded the simple additive models. The gain from the use of nonlinear procedures was usually lost on cross-validation. Goldberg's findings thus provide evidence that configural actuarial models result in greater shrinkage in the correlation between predictor variables and criterion upon cross-validation than simple linear combinations.

Sufficient background material has now been presented on the linear regression model and the configural approach. The reports that follow offer illustrations of prediction from taxonomic classes using multiple regression as well as cross-product terms. Other studies compare the predictive efficiency of subgroups alone with that of pooled data.

Recent work has carried these findings with regard to nonlinear relations still further to linear regression coefficients. An early paper by Wilks (1938) proved, under a set of reasonably general conditions, that with a sufficiently large number of intercorrelated predictor variables almost any combination will yield the same prediction. Einhorn and Hogarth (1975) showed that *equal* regression weights would be a reasonable choice to weight a set of positively correlated predictors. Wainer (1976) subsequently showed that under general circumstances coefficients in multiple-regression

models can be replaced by equal weights with almost no loss in accuracy on the original data sample. Moreover, these equal weights will have greater robustness than least-squares regression coefficients. Thus when interest is mainly in prediction these findings suggest that the researcher orient all predictor variables in the correct direction relative to the criterion, scale all variables into standardized form, and add up the unit-weighted variables.

Biodata or Life-History Studies

Owens (1968, 1971), following Toops, proposed that subgroups of individuals be constructed on the basis of biodata or life-history variables. This procedure assumes that the best predictor of future behavior is past behavior. A corollary of this axiom is that those who have behaved in a similar fashion in the past (members of a coherent subgroup) will behave in similar ways in the future. Owens (1971) and Schoenfeldt (1970) administered an identical biodata form of 118 items to approximately 1,000 male college freshmen and 900 female college freshmen at a southeastern university. The items, factored separately by sex, resulted in the identification of 15 interpretable factors in the female data and 13 in the male data. Of these, 10 factors were very similar in both sexes; there were also five unique female factors and three male factors. Each subject was profiled on the factor dimensions. A matrix of distances between profiles was computed and analyzed by Ward's hierarchical procedure.

Some of the factored dimensions of the male profiles were warmth of parental relationship, academic achievement, social introversion, independence/dominance, athletic interest, socioeconomic status, and parental control. Schoenfeldt (1970), using all 23 male subgroups, demonstrated that they differed significantly in academic performance as evidenced by honors, dropouts, and number of extracurricular activities. In another study Pinto (1970) administered a comprehensive biodata form plus certain standardized measures to 2,060 salesmen. For the analysis, the overall sample was split into subsamples *A* and *B*. The subjects of subsample *A* were subgrouped by Ward's procedure; the subjects of subsample *B* were assigned to those subgroups with only a 10 percent loss because of

poor fit. Termination reports constituted the dependent variable, which was predicted by means of several procedures. Use of measures other than biodata to predict the criterion showed no results. However, the biodata subgroups correlated significantly with the criterion in both samples.

Ruda (1970) had 458 executives of a large oil company complete a 247-item personal history form. Top-level executives also ranked their subordinates' performances. All subjects were subgrouped by Ward's procedure, and 13 subgroups were identified. Distance from superior's to subordinate's subgroup (a measure of resemblance) was found to be negatively related to subordinate's rated performance. In other words, similarity between rater and ratee in biodata profile predicted rated performance. Owens (1976) gives a full report on the advantages of background data.

Another life-history study has been reported by Feild, Lissitz, and Schoenfeldt (1975). Their study was designed to determine whether group information was superior to individual predictor information and whether the prediction of criterion measures by individual data could be enhanced by the addition of group data. A 375-item biodata questionnaire (Biographical Information Blank) was administered to 1,037 freshmen males and 897 freshmen females at a large southeastern university. A principal-axis factor analysis of the BIB, completed separately for each sex group, yielded 13 factor scores for males and 15 for females. Life-history profiles based on the standardized factor scores were computed for each person. Next Ward's hierarchical procedure was applied separately to the distances between members of each sex group to generate homogeneous subgroups. In the male sample 23 subgroups (mean size of 33) were found; in the female sample 15 subgroups (mean size of 44) were identified.

The criterion information was based on college experiences reported in the College Experience Inventory (CEI) given to 743 university seniors. A factor analysis yielded 12 interpretable factors. The predictive study was actually based on 509 university seniors on whom BIB and CEI data had been collected over 4 years. The 509 subjects were allocated to 19 male and 13 female biodata subgroups. Canonical correlations were used to determine the extent of the

relationship between individual information (BIB factor scores) and biodata subgroup information. Moreover, multiple correlations (R) were computed between the predictor sets and each of the 12 CEI factors. For subgroup information to be useful as a predictor, it must increase the multiple R when added to individual information. The findings indicated, in both men and women, that the biodata factor scores (the predictors) were more significantly related to the CEI factor than to the biodata subgroup information. However, the *addition* of subgroup information to individual information significantly enhanced the multiple R^2 in predicting two criteria in men. It also increased the predictability of two criteria in females. The reason is that the grouping dimensions act as moderators between predictors and criteria. In using a subgroup, it is assumed that there is no meaningful within-group variance. To the extent that subgroup means differ from the grand mean and the subgroup variance is less than the grand variance, subgroup information will contribute to prediction. In summary, then, the prediction of several criteria by individual information was significantly improved by the addition of subgroup information.

Subgroups vs. Pooled Samples

Frank (1980) compared an actuarial prediction model with the linear model; both were developed on an identical set of variables in the same sample. There were four phases of the research: (1) development of homogeneous subgroups, (2) validation and cross-validation of predictions made from subgroup membership, (3) validation and cross-validation of predictions made from the linear model, and (4) comparison of the actuarial and linear models.

The subjects were 5,798 employees tested in a managerial assessment program. The sample was randomly divided into validation and cross-validation samples. Two criteria relevant to organizational success were developed: employment status and job performance. Regarding employment status, employees were assigned to a category containing those who were with the company or had been retired or to a second category including those who had left or had been terminated. For the second criterion, job performance, a

management assessment test battery furnished analysis variables. The 13 scores included BIB scores, managerial judgment scores, and temperament survey scores.

For the *actuarial model,* employees in one sample were grouped according to similarity with respect to the 13 assessment score profiles by way of Ward's hierarchical clustering procedure. The number of subgroups retained was based on a criterion similar to Mojena's Rule One described in an earlier chapter. The 12 subgroups retained were next subjected to a *k*-means cluster analysis. For cross-validation the second sample of 2,899 subjects was assigned to one of the subgroups in the validation sample by means of a minimum-distance qualifier. The cross-validation profiles replicated the validation sample profiles very closely.

For the *linear model,* a discriminant analysis was performed on employment status using the 13 analysis variables as predictors. In the cross-validation sample the discriminant weights in the first sample were used to develop a predicted employment status. For the regression analysis the validation sample's performance scores were used to predict performance scores for each member. In the cross-validation sample the predicted performance score weights were applied to the 13 variable scores.

The actuarial model was tested by a chi-square test of association between the 12 employee subgroups and employment status. Although the tests were significant, knowledge of subgroup membership was not useful for predicting employment status. In testing the value of subgroup membership for predicting job performance, the chi squares were highly significant. Knowing subgroup membership resulted in a 26 and 25 percent reduction in the probability of error in predicting job performance. Hit rates of 64 and 63 percent indicated that subgroup membership made for 12 to 13 percent more correct prediction than by base rate alone.

The discriminant analysis for employment status, as in the actuarial model, indicated little gain over base rate. Multiple correlation of 0.45 resulted from the regression of job performance on the 13 analysis variables; the cross-validation multiple correlation too was 0.45. Calculated hit rates of 64 and 68 percent indicate that regression analysis of job performance was worthwhile.

The findings showed that the subgroups were significantly related to one external criterion moreover, and, that the actuarial and linear models were equally effective in predicting these criteria. Frank argues, however, that the actuarial model is very efficient. Once a person's subgroup membership has been determined, predictions can be made to many external criteria. The linear model, in contrast, requires the generation of a separate regression or discriminant equation for each criterion and subsequent development of a classification table for interpretation. Second, and perhaps more important, the actuarial model promotes understanding. Being a member of one subgroup rather than another has a unique meaning as represented by the subgroup profile of scores. A descriptive taxonomy based on the subgroups also becomes available.

Configural vs. Linear Prediction

As described earlier, Feild and colleagues compared the utility of homogeneous subgroups with that of individual information in prediction. They found that prediction by individual information was significantly improved by the addition of subgroup information. Frank (1980) found that the linear and actuarial models were essentially equivalent when judged by predictive accuracy. Recently Micucci (1980) used *linear multiple regression* (LMR) to investigate linear and configural relationships between the MMPI scales and external criteria. He followed one variant of LMR developed by Cohen and Cohen (1975) called *hierarchical regression.* In this procedure independent variables are entered into the prediction equation in steps. The order of entry of the variables into the equation can be specified in advance on the basis of causal priority or importance. The difference between the squared multiple correlations at each step gives the amount of additional variance explained by the addition of the variable just entered to those already in the equation. When the independent variables are categorical rather than continuous they can be included by use of the dummy variable coding described in an earlier chapter. Another feature of Micucci's study was the evaluation of the amount of variance in a criterion variable accounted for by configural interactions among predictor variables. As indicated

earlier, configural or nonlinear relationships among independent variables can be expressed as cross-product terms.

Four sets of independent variables were studied: MMPI scale scores, cross-product configural terms, and dummy variables to handle both the Marks/Seeman/Haller (1974) code types and Gilberstadt's *P* code types (1970). Since *P* code types were available only for the Gilberstadt/Duker (1965) sample, these variables were considered for this sample only. Data from one sample came from Gilberstadt's *P* code atlas; the other sample consisted of female profiles taken from the Marks/Seeman code book (1963). The dependent variables included diagnostic classification, symptom clusters, syndrome clusters, and outcome. The Gilberstadt data were used to test validity; the Marks/Seeman data (using simplified code typing) served as the replication sample. Now let us examine some of the findings. First, the Marks/Seeman/Haller (MSH) codes, when added as dummy variables in the LMR equations, did not contribute to the prediction of the dependent variables to a significant degree in the Gilberstadt sample. Second, when the MSH code designations were the only independent variables in the LMR equations, the average amount of criterion variance explained over all dependent variables was quite small. And third, for the Marks/Seeman sample the MSH codes performed somewhat better than in the Gilberstadt sample but still did not exceed the predictive effectiveness of the 13 MMPI scales in linear combination.

In the Marks/Seeman sample, addition of the MSH codes to an LMR equation containing the 13 MMPI scales yielded only a small increase in criterion variance explained. Classification by Gilberstadt's *P* code typing resulted in better prediction of criteria than classification by MSH code typing but did not exceed the prediction possible by optimal linear combination of the 13 scales. The addition of cross-product (interaction) terms in the Gilberstadt sample resulted in significant increases in the amount of criterion variance explained. These findings were not replicated in the Marks/Seeman sample, however.

The general conclusion is similar to that of Goldberg. The use of subgroup information in MMPI prediction is not supported in this research. By far the greatest amount of criterion variance is explained by the linear combination of the 13 scales.

Pooling vs. Subgrouping

Gross (1973) sought to compare the predictive accuracy of two different prediction strategies. The first strategy was the usual procedure of computing a least-squares regression equation of the dependent variable Y on p independent variables. Since the entire sample is used, this strategy is referred to as *pooling*. The second strategy is a subgrouping procedure whereby one partitions a sample of size N into $g \geq 2$ subgroups and then computes a separate regression equation and multiple correlation within each group. Gross defined predictive accuracy in terms of expected squared cross-validity (ESCU), expected mean square error (EMSE), and gain selection (GS). He cites six studies in which, after partitioning a sample into g subgroups and performing a separate regression analysis within each group, one or more subgroups could be identified having a larger multiple correlation than the R for the total sample. One question is whether this type of finding will be maintained under cross-validation. The problem is to select a certain proportion of the population such that the average value of the dependent variable within the selected group is maximized.

The mixture model assumed by Gross views the population of subjects as a union of disjoint subpopulations. Published data from two studies that employed both a subgrouping and a pooling prediction approach were used to compare the two strategies. These strategies provided data for seven different subgrouping analyses. In 23 of 34 subgroups, regression equations had multiple correlations higher than the corresponding total sample regression equations. Fourteen of the subgroup equations showed higher ESCU than the corresponding total sample equations. Furthermore 16 subgroup equations had smaller EMSE than the corresponding total sample equations.

Gross concludes that a subgrouping analysis definitely shows promise as a model for improving the accuracy of prediction. He found a clear advantage for the subgrouping approach in four of the six studies. The advantage is seen only when the selection ratio is 0.50 or less, however; when more than half the population is to be selected, the advantage is small. For the advantage to be realized, the sample size must be more than 100. With small samples, the advan-

tage of the subgrouping model is likely to be lost due to greater sampling error present in the subgroup regression weights compared to the total sample weights.

Summary

This chapter has covered what has become known as prediction from a taxonomic class. Persons or other entities are classified in terms of patterns of test or life-history variables selected as broadly relevant to the types of behavior one wishes to predict. Actuarial or conventional criterion forms of data can then be collected, tabulated, and tested for their ability to distinguish among the subgroups. Biodata or life-history information has shown particular value for such studies in psychology. Since the findings from the various methodological studies reviewed here are by no means consistent, the need for cross-validation and for samples of good size must be emphasized. At the same time Gross's conclusion that subgrouping analysis definitely shows promise as a model for improving accuracy of prediction should provide encouragement for further studies.

CHAPTER 10

Overlapping Clusters

Most methods of clustering, hierarchical as well as nonhierarchical, are constrained to yield mutually exclusive and exhaustive categories. The objective of this chapter is to describe a new clustering model—*additive clustering*—that allows for the generation of overlapping clusters. The assumption made in the model is that the similarity of two objects is a simple additive function of underlying weights associated with whatever properties are shared by both objects. The method of additive clustering (ADCLUS) was developed by Shepard and Arabie (1979). The procedure may be used in studies involving entities (subjects as well as products) that belong to more than one subgroup or cluster simultaneously. Because of computational difficulties the algorithm ADCLUS did not lead to generally successful solutions. The procedure to be described, mathematical programming clustering (MAPCLUS), was devised by Arabie and Carroll (1980) for fitting the ADCLUS model. Some diverse illustrations will be presented.

Representations of Similarities

Spatial models of two-way and three-way multidimensional scaling for the discovery and representation of structures in data are

well known. In a geometric representation of factor analysis that corresponds closely to multidimensional scaling, points represent variables and dimensions represent entities. Both models assume that measures of similarity or dissimilarity relate in simple ways to distance in a metric space. When one applies a clustering procedure to a matrix of proximities, one is fitting a kind of nonspatial model. The model may be set-theoretic if stated in terms of discrete qualities rather than dimensional attributes. Or the model may be a tree structure if fitted to a hierarchical clustering. Carroll (1976) has presented numerous examples of spatial, nonspatial, and hybrid models. He proposes that trees be viewed as intermediate between multidimensional scaling and clustering.

J. B. Kruskal (1977), like Carroll, emphasizes the complementary nature of dimensional scaling and clustering. Clustering and dimensional scaling may give equally accurate representation of a set of data in actual practice. But he observes that in hierarchical clustering small clusters fit well and provide meaningful groupings whereas large clusters high in rank fit poorly and are less meaningful except in the family trees of biology. In multidimensional scaling local features fit poorly since they reflect small dissimilarities while global positions reflect large dissimilarities. When large dissimilarities are deleted the structure is badly damaged. Since dimensional scaling and clustering are sensitive to complementary aspects of a data set Kruskal suggests that both methods be used. In fact it has become common to portray clusters in a spatial representation by drawing loops around the clusters.

An alternative to continuous spatial models of multidimensional scaling and factor analysis, and the models of discrete clusters, has been developed by Shepard and Arabie (1979). Their additive clustering model falls between multidimensional scaling and mutually exclusive discrete models. Similarities are represented as combinations of discrete overlapping properties. The method provides an explicit assignment of N entities under study to discrete subsets that can overlap. The model subsumes hierarchical clustering as a special case, and the procedure can be considered a discrete form of principal-component analysis.

Overlapping vs. Nonoverlapping Clusters

Although most cluster techniques are designed to recover discrete clusters without overlap, some nonhierarchical partitioning procedures yield clusters with a slight amount of overlap. Explicitly discrete, categorical representations of structure in similarity data are usually achieved by means of one of the popular hierarchical clustering methods. These procedures all yield a tree structure in which at any single level of the hierarchy none of the clusters overlap since entities are allocated into mutually exclusive and exhaustive subsets. From the viewpoint of the social scientist such methods yield an excessive number of clusters unless a developmental sequence is sought. To reduce the number of subsets to an optimal number reflecting actual structure in the data, several procedures can be applied. Typically a horizontal slice of the hierarchy is chosen; Baker and Hubert (1975) have proposed criteria for such a procedure. Another method is to apply Mojena's Rule One described in an earlier chapter.

A number of techniques for deriving overlapping clusters have been available for some time, but none have gained popularity. Among these are the B_k procedure devised by Jardine and Sibson (1971) and the Peay method (1975). One disadvantage of these methods is that they produce too many clusters, which may call for the imposition of arbitrary constraints. A second handicap is that if the clusters are too inclusive, embedding them in a spatial solution such as multidimensional scaling is generally difficult. But it is not difficult to find examples in which clusters overlap.

Shepard and Arabie (1979) point out that in applying hierarchical clustering methods the constraints require aunt to be grouped with uncle and niece with nephew on the basis of generation. Then it is no longer possible to join aunt with niece or uncle with nephew on the basis of sex. In other words, two cross-cutting classifications cannot be simultaneously accommodated within a hierarchical arrangement. Similarly father and mother are classified together in contrast to son and daughter on one basis, while excluding a fusion of mother and daughter and father with son on the basis of sex.

Other illustrations may be found in analyzing cliques, lobbies, and political factions when these interlock. A member may belong to several cliques or political factions rather than one.

In psychiatric classification it has long been known that certain diagnostic classes overlap. Psychiatric patients may belong to more than one symptom group despite the all-or-none disease model adopted—for example, some cases may be classifiable both as paranoids and as schizophrenics. With respect to the personality disorders set forth in DSM-III (American Psychiatric Association, 1980), patients may belong simultaneously to the narcissistic, the histrionic, and the borderline (impulsive). Marketing analysts allocate brands as well as consumers to only one cluster although brands often compete against each other and consumers properly belong to several subgroups (Arabie and others, 1981). Certainly a brand can compete in more than one cluster of products—for example, Bayer's aspirin may compete with Bufferin to relieve pain or fever and compete with Anacin and Excedrin to relieve arthritic complaints. Analogously consumers may want to prevent bad breath *and* tooth decay with a toothpaste. Studies of professional organizations bring out comparable problems. Individuals often belong to several specialty groups or divisions of an organization. The problem is to isolate overlapping groups as a guide to membership interests.

The Additive Model

The purpose of the additive clustering model is to represent data as overlapping subsets (clusters). Although the model can be applied to any data classed as proximities, including dissimilarities as well as similarities, it is usually applied to similarities and thus the discussion will be confined to this form. Assume that N entities are to be clustered with the input data of $N(N-1)/2 = m$ entries arranged into a two-way symmetric (or symmetricized) proximity matrix. Each of the hypothesized overlapping subsets of entities is assumed to correspond to a discrete property shared by all and only those entities within a subset. It is further assumed that each property contributes a fixed increment to the similarity between any two entities sharing that property (Shepard and Arabie, 1979). Each contribution is independent of the contribution of all other properties.

The basic equation of the additive model is thus

$$\hat{s}_{ij} = \sum_{k=1}^{n} w_k p_{ik} p_{jk} \tag{10-1}$$

where $\hat{s}_{ij}$ is the theoretically constructed similarity between entities i and j, w_k is the nonnegative weight representing the salience (prominence) of the property, and

$$p_{ik} = \begin{cases} 1 & \text{if entity } i \text{ has property } k \\ 0 & \text{otherwise} \end{cases}$$

The product $p_{ik}p_{jk}$ is unity if both entities belong to subset k; it is zero if either falls outside that subset. Since w_k times unity is still w_k, $\hat{s}_{ij}$ is simply the sum of the weights.

Expressed in matrix notation to include all similarities, Equation (10-1) becomes

$$\hat{\mathbf{S}} = \mathbf{PWP'} \tag{10-2}$$

where $\hat{\mathbf{S}}$ is an $N \times N$ symmetric matrix of reconstructed similarities, $\mathbf{W}$ is an $m \times m$ diagonal matrix with weights $\mathbf{W}$ $(k = 1, 2, 3, \ldots, m)$ in the principal diagonal, $\mathbf{P}$ is the $N \times m$ rectangular matrix of binary values p_{ik}, and $\mathbf{P'}$ is the transpose of matrix $\mathbf{P}$. Each column of $\mathbf{P}$ represents one of the m subsets. The entries $\hat{s}_{ij}$ in the computed matrix are to be fitted to the entries s_{ij} in the empirically obtained matrix $\mathbf{S}$.

Shepard and Arabie (1979) require that the mth subset consist of a column of 1's to provide an additive constant to Equation (10-1). This requirement is needed to assess the variance accounted for by the theoretically reconstructed similarities $\hat{s}_{ij}$. Arabie and Carroll (1980) prefer to express Equation (10-2) as

$$\hat{\mathbf{S}} = \mathbf{PWP'} + \mathbf{C} \tag{10-3}$$

where $\mathbf{C}$ is an $N \times N$ matrix having zeros in the principal diagonal and the fitted additive constant c in all the remaining entries.

As stated earlier, Arabie and Carroll devised MAPCLUS to replace the original algorithm. A major difference between the two

programs is that a fairly small number of subsets are prespecified by the user at the beginning of the analysis. The detailed description is given in Arabie and Carroll's report and a manual is available from Bell Laboratories (Murray Hill, New Jersey 07974). Since the process is quite complex, only a broad sketch can be given here. The **P** matrix is initially considered to have continuously varying p_{ik} whereas **W** is initially all zeros. The overall algorithm constitutes a *mathematical programming* approach to clustering. A discrete solution is approached by a sequence of increasingly close continuous approximations.

The capability of the current version of the MAPCLUS algorithm is as follows. The maximum number of stimuli (entities) allowed is 30; the maximum number of subsets including the complete subset is 25. A suggested procedure is to (1) cluster a sample of any size by any clustering algorithm (say Ward's) and select a set smaller than 30; (2) develop a matrix of interentity similarities based on the average size of the pseudoentities; (3) submit the data matrix to MAPCLUS algorithm. Multidimensional scaling can then be applied to the data to gain an improved interpretation of the results.

Applications

The value of overlapping clustering can be illustrated on data for 15 breakfast foods collected by Green and Rao (1972). The data consisted of similarity judgments for each of 15 breakfast foods given by 42 respondents (21 male students and their wives). Each subject's judgments were converted to a matrix of dissimilarities between the $15(14)/2 = 105$ pairs of food items. MAPCLUS was applied to the data set and varying numbers of subsets ranging from 10 to 4 were specified. As Arabie and others (1981) point out in their analysis, there is a trade-off between interpretability and goodness of fit as in factor analysis and multidimensional scaling. The four-cluster solution was not interpretable, while clusters in the 6 to 10 range were too inclusive. Five clusters, enclosed by contours, were embedded in a two-dimensional scaling solution found by Green and Rao (1972). These clusters are presented in Figure 10-1; the stimuli are listed in Table 10-1.

Figure 10-1. Overlapping Clusters of Breakfast Food Items.

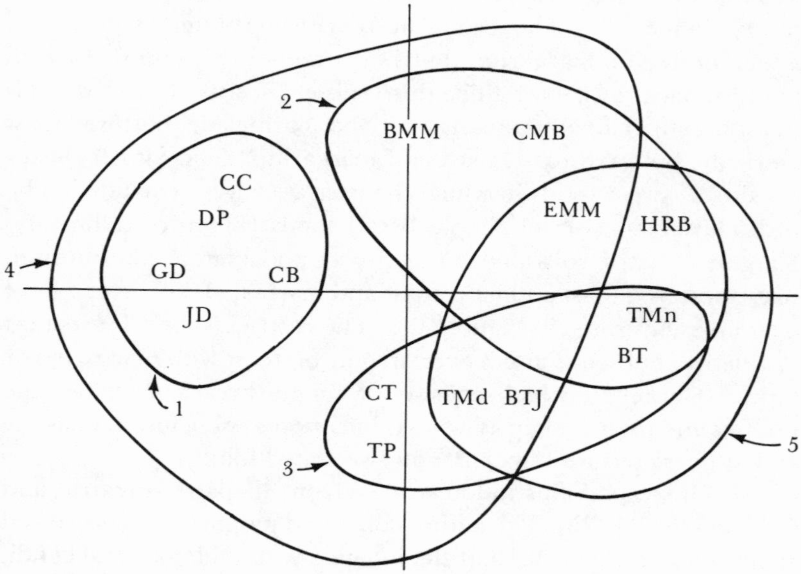

Source: Arabie and others, 1981. Used by permission.

Table 10-1. Breakfast Food Items Used by Green and Rao.

Food Item	*Plotting Code Used in Figures*
1. Toast pop-up	TP
2. Buttered toast	BT
3. English muffin and margarine	EMM
4. Jelly doughnut	JD
5. Cinnamon toast	CT
6. Blueberry muffin and margarine	BMM
7. Hard rolls and butter	HRB
8. Toast and marmalade	TMd
9. Buttered toast and jelly	BTJ
10. Toast and margarine	TMn
11. Cinnamon bun	CB
12. Danish pastry	DP
13. Glazed doughnut	GD
14. Coffee cake	CC
15. Corn muffin and butter	CMB

The highest cluster (ranked on weight) consists of pastry items (jelly doughnut, glazed doughnut, cinnamon bun, Danish pastry, coffee cake). The second cluster consists of items spread with either butter or margarine (blueberry and corn muffins, English muffins, hard rolls, toast). The third subset consists of toasted foods (toast pop-up, cinnamon toast, toast and marmalade, buttered toast and jelly, buttered toast, toast and margarine). The fourth cluster consists of sweet foods in which the nonsweet were excluded. The fifth cluster consists of simple bread foods (English muffins and margarine, hard rolls and butter, toast and margarine, buttered toast, buttered toast and jelly, toast and marmalade). The extent of overlap is illustrated in Figure 10-1. The pastries (subset 1) are most distinctive. Subsets 2 and 3 overlap only on toast with margarine or butter. Cluster 5, which consists of simple breads, excludes the muffins and toast pop-up as well as cinnamon toast. Cluster 4 has the widest overlap since it includes all sweetened foods.

The second illustration is taken from Shepard, Kilpatric, and Cunningham (1975). The authors obtained judgments of perceived similarities between all 10 of the integers 0–9 under several conditions. The illustration reported here involved judgments made with respect to the abstract concepts of the numbers themselves. The 10 recovered subsets accounted for 83.1 percent of the variance of the judged similarities. The subsets listed in order of weight were as follows: (1) powers of 2 (2, 4, 8), (2) large numbers (6, 7, 8, 9), (3) middle numbers (3, 4, 5, 6), (4) small nonzero numbers (1, 2, 3), (5) multiples of 3 (3, 6, 9), (6) additive and multiplicative identities (0, 1), (7) odd numbers (1, 3, 5, 7, 9), (8) moderately large numbers (5, 6, 7), (9) small numbers (0, 1, 2), (10) smallish numbers (0, 1, 2, 3, 4).

A third illustration involves a representation of data taken from Miller and Nicely's (1955) experimental investigation of confusion of 16 English consonant phonemes. Psychologists have for many years sought a preferred set of discrete underlying features of consonant phonemes (speech sounds). Shepard and Arabie's ADCLUS solution for confusion between 16 phonemes yielded 16 clusters; Arabie and Carroll (1980), using MAPCLUS, found a solution with 8 clusters. The largest weighted subsets consisted of (1) front unvoiced fricatives (f, Θ), (2) front voiced fricatives (v), (3) back voiced stops (d, g), (4) unvoiced stops (p, t, k), (5) front voiced

consonants (*b*, *v*), (6) unvoiced stops, omitting *t*(*p*, *k*), (7) voiced consonants (*b*, *d*, *g*, *z*), and (8) unvoiced consonants, omitting *t*(*p*, *k*, *f*, θ).

Summary

This chapter has presented an alternative to the spatial models of multidimensional scaling and factor analysis and to the mutually exclusive discrete models of cluster analysis. The illustrations given here argue for a model of overlapping clusters. Some of the applications we considered were based on the ADCLUS algorithm and others on MAPCLUS developed by Arabie and Carroll to replace the original program. It can be concluded that the additive model of overlapping clusters satisfies a real need in certain cluster problems. The hierarchical model itself is a special case of this general conceptual scheme.

APPENDIX A

Computer Programs for Performing Cluster Analysis

This appendix is intended to provide users of cluster analysis with a list of available programs and their sources. Information regarding the advantages and disadvantages of these popular programs is also presented so that users can choose the one appropriate to their needs. Additional details may be found in Blashfield's (1976b) consumer reports on cluster analysis software. A sample of scientists was sent a questionnaire concerning seven statistical packages that include cluster programs. They were asked which programs they were familiar with and which they used most frequently. Subsequently Blashfield and Aldenderfer (1978) reviewed the literature published since 1964 in which cluster analysis was used for data analysis.

Sixteen criteria were applied to evaluate the various cluster programs. Of these criteria the two judged most important by users were the versatility of the clustering options and the clarity of the user's manual. Versatility was deemed important because users want a sufficient range of options that allow for a choice and for the exploration of unfamiliar approaches. Other criteria included usefulness of the graphics, readability of the output, provision for basic descriptive statistics, cost of executing a data analysis, and options for handling missing data.

Cluster-analytic techniques may be categorized into three

main groups: hierarchical agglomerative methods, methods of iterative partitioning, and miscellaneous methods (obverse factor analysis, clumping, minimum spanning tree, and others). Blashfield found three main sources of cluster programs: statistical packages, programs that perform only one kind of cluster analysis, and programs presented in books. The principal statistical packages to be described here are BCTRY, BMDP, CLUSTAN, NT-SYS, OSIRIS, and SAS. Programs presenting only one method of clustering are HGROUP and HICLUS. The books with program collections are by Anderberg (1973) and Hartigan (1975).

Blashfield (1976b) reports that four hierarchical methods are clearly preferred as clustering techniques: single linkage, complete linkage, average linkage (UPGMA), and minimum variances (Ward). The best-known partitioning procedures are *k* means, CLUS (Friedman and Rubin), and Wolfe's NORMIX and NORMAP.

Programs Containing Cluster Analysis Methods

General Statistical Packages

BCTRY. The routines available on BCTRY were developed by R. C. Tryon and D. Bailey. The two cluster analysis routines are EUCO, which performs a *Q* analysis, and OTYPE, an iterative partitioning method. The remaining routines concern dimensional analysis. Other grouping procedures and some cluster theory are presented in Tryon and Bailey's *Cluster Analysis* (1970). For an updated program write to Robert B. Dean, 1629 Columbia Road, N.W., Washington, D.C. 20009.

BMDP (1981). The BMDP series represents a general statistical package with routines for performing simple data analysis, correlation, regression, factor analysis, and canonical correlation analysis. There are four programs that form clusters of variables or cases or both: P1M, P2M, PKM, and P3M. P1M forms clusters of variables by following an average-linkage hierarchical agglomerative procedure. P2M joins cases or clusters of cases in a stepwise process until all cases are combined into one cluster. The linkage rules for joining clusters include single linkage, centroid linkage, or the *k*th nearest

neighbor. It prints either a vertical or horizontal tree diagram. PKM represents a *k*-means clustering technique and establishes a fixed number of homogeneous groups of cases using distances. The program begins with user-specified clusters or with all cases in one cluster and splits one cluster into two at each step. When the specified number of clusters is reached, cases are iteratively reallocated into the clusters whose center is closest. The output includes a scatterplot of the orthogonal projection of cases into the plane defined by the centers of the three most popular clusters. P3M forms clusters of both the cases and the variables in a data matrix simultaneously. This program prints a block diagram to describe the blocks identified and tree diagrams, one for the cases and one for the variables. *BMDP Statistical Software 1981* (W. J. Dixon, editor) is published by the University of California Press.

NT-SYS. The NT-SYS package of multivariate statistical techniques for biologists includes programs for factor analysis, multidimensional scaling, and cluster analysis. Five different hierarchical agglomerative procedures are included as well as 21 similarity/dissimilarity measures. All the popular hierarchical methods are included except minimum variance. The procedure for the on-line plotting of hierarchical trees is especially good. The program package may be obtained by writing to F. J. Rohlf, Department of Ecology and Evolution, State University of New York at Stony Brook. The acronym NT-SYS stands for *Numerical Taxonomy System of Multivariate Statistical Programs.*

OSIRIS. OSIRIS, a general statistical package intended for use by social scientists, offers routines for data description and tabulation. One routine, called CLUSTER, contains a procedure for overlapping clusters of variables. The other procedure, HICLSTR, is hierarchical and is also for clustering variables; with suitable modification, however, cases can be clustered. OSIRIS is available from the Institute for Social Research, University of Michigan, Ann Arbor.

SAS. SAS is a general statistical package that includes CLUSTER, a subroutine for Johnson's hierarchical agglomerative complete-linkage analysis. It also prints out a cluster map. The program is available in *SAS User's Guide* (1979 edition) edited by A. J. Bar, J. H. Goodnight, and J. P. Sall. The guide is obtainable from SAS Institute, Inc., P.O. Box 10066, Raleigh, NC 27605.

Cluster Analysis Package. CLUSTAN represents a collection of cluster analysis routines assembled by David Wishart. Included are seven routines for hierarchical agglomerative as well as hierarchical divisive analysis; other algorithms are devoted to iterative partitioning, mode searching, and minimum spanning tree methods. Some 38 similarity/dissimilarity measures can be computed. Graphics (trees) are handled through off-line devices. The package may be characterized as highly versatile because of the large number of clustering and similarity measurement options. The *CLUSTAN User Manual* (3rd ed., 1978) may be obtained from the Program Library Unit, Edinburgh University, 18 Buccleuch Place, Edinburgh EH8 9LN, Scotland.

Programs with One Type of Cluster Analysis

BUILDUP. The BUILDUP algorithm follows a nonhierarchical average-linkage procedure that generates successive clusters. Inclusion criteria for membership in a cluster and exclusion criteria for removing outliers must be specified in advance. The inclusion-exclusion limits are typically set at values where a coefficient of correlation based on k variates is significant at $p = 0.025$ and $p = 0.05$. The source is M. Lorr, Department of Psychology, Catholic University, Washington, D.C. 20064.

CLUS. The CLUS program is based on a report by H. P. Friedman and R. J. Rubin (1967). The program emphasizes iterative partitioning procedures to optimize certain functions. It is relatively complex and expensive and calls for considerable sophistication in computer analysis. It is available through the IBM SHARE system.

HGROUP. The HGROUP program follows Ward's (1963) minimum-variance method of hierarchical clustering. It is a routine contained in Veldman (1967).

HOWD. The HOWD program uses a hierarchical divisive procedure to determine the number of clusters. Then all the entities are reallocated successively to the cluster with the closest centroid. Copies may be obtained from F. J. Carmone, Jr., Marketing Science Institute, Philadelphia, PA.

HICLUS. The HICLUS program, like the one in SAS, is based on Johnson (1967). The program uses single-linkage and complete-

linkage hierarchical agglomerative methods to analyze a distance matrix.

ISODATA. ISODATA follows a complex *k*-means program developed at Stanford Research Institute. A critical question is when to stop iteration. The program emphasizes several approaches to splitting and merging clusters. Because of the need for interactive graphics, one version is incorporated into a PROMENADE system. The source of ISODATA is the Stanford Research Institute, 333 Ravenswood Avenue, Menlo Park, CA.

MIKCA. MIKCA is an iterative *k*-means cluster analysis program that seeks to optimize one of four user-chosen criteria such as the determinant of **W** (pooled within-cluster covariance matrix). The program is reported by McRae (1971).

Milligan. Two computer programs for a two-stage clustering algorithm with robust recovery characteristics are described by Milligan and Sokal (1980). In stage 1, a group-average hierarchical algorithm generates cluster centroids that are then used as seed points for stage 2, Jancey's *k*-means nonhierarchical algorithm. Program listings are available from G. W. Milligan, Faculty of Management Sciences, Ohio State University, Columbus, OH 43210.

NORMIX. The NORMIX program (described in Wolfe, 1970) deals with the general case while NORMAP makes simpler assumptions. The method of maximum likelihood is applied to estimate mixture proportions, means, and covariance. Objects are assigned to the group for which their probability is greatest. The method calls for large sets of data and ample computer time. The program is available from Naval Personnel and Training Research Laboratory, San Diego, CA 92132.

Collections of Cluster Programs in Books

Anderberg. Anderberg (1973) has a number of FORTRAN subroutines that contain seven varieties of hierarchical agglomerative clustering and three varieties of *k*-means partitioning. Also included are subroutines for scale conversion and association measures.

Hartigan. A wide range of approaches to clustering techniques is offered in Hartigan (1975). Included are hierarchical

agglomerative and divisive clustering methods and several partitioning procedures such as *k*-means and normal mixture algorithms. Each chapter presents a step-by-step description of a method along with FORTRAN notation and closes with the program.

Programs Using Q Analysis

LTA. Linear typal analysis (LTA) is a factor analysis program for identifying clusters of individuals. Person factors are conceived as ideal types; observed profiles are viewed as weighted averages of pure type profiles. The theory and rationale as well as several illustrations are presented. By means of a unique feature—the addition of a constant to all mean-corrected scores—bipolar person factors are avoided. The program is given in Overall and Klett (1972).

MPA. Modal profile analysis (MPA) is designed for identification and cross-validation of relatively homogeneous subgroups within a dimensional or ordination space. The computations are based on a sequential application of singular value decomposition. The program may be obtained from H. A. Skinner and H. Lee, Addiction Research Foundation, 33 Russell Street, Toronto, Ontario M5S 2S1. The computer program is reported in Skinner and Lee (1980).

Programs for Multidimensional Scaling for Ordination

M-D-SCAL. This two-way approach to MDS was initiated by Shepard and developed by J. B. Kruskal. The latest version available from Bell Laboratories is M-D-SCAL-5. Write to Computing Information Services, Attn: Mrs. Irma Biren, Bell Laboratories, Murray Hill, NJ 07974.

KYST. Young, Kruskal, and Seery combined features of a metric program called TORSCA with M-D-SCAL. This program is also available from Bell Laboratories.

INDSCAL. INDSCAL, a three-way approach to MDS, was designed by Carroll and Chang (1970) to carry out individual differences scaling. Widely used because it is powerful, it is available from Bell Laboratories.

ALSCAL. The ALSCAL program, created by Takane, Young, and DeLeeuw, generalizes INDSCAL somewhat and has some computational advantages. The program can handle either two-way or three-way data. It is available from the Thurstone Psychometric Laboratory, Attn: Forrest Young, University of North Carolina, Chapel Hill, NC 27514.

SSA-I to IV. Smallest-space analysis is an approach due to L. Guttman. The procedure resembles Torgerson's and Shepard's procedure. The program is available from J. C. Lingoes, 1000A North University Bldg., University of Michigan, Ann Arbor, MI 48104.

Comments Regarding Programs

Several dimensions of variability characterize the various cluster methods. Bailey (1974) lists 12 criteria; Sneath and Sokol (1973) suggest 8 options; Everitt (1974) suggests 5 classes. The scheme presented in Chapter Five classifies the methods into single-level nonhierarchical and multilevel hierarchical. These two dimensions in turn can be subdivided into linkage forms such as single, complete, and average. A third basis for classification is the similarity and dissimilarity measures used in the linkage process. The two broadest classes are distance measure and measures of angular separation (correlations and association coefficients). A fourth set of criteria divides the single-level techniques into iterative partitioning, successive single-cluster methods, and density-seeking or mode-seeking techniques. The multilevel methods can be classified as agglomerative or divisive.

Similarity Indices. One question of concern to the user is the number of similarity/dissimilarity measure options available in a cluster analysis package or program. CLUSTAN offers 38 coefficients covering all types for continuous as well as binary categorical data; NT-SYS contains 21 measures; Anderberg has 15 options; BMDP offers 9; Hartigan has 4.

Multilevel vs. Single Rank. Both hierarchical and iterative partitioning procedures may be found in Anderberg, BMDP, CLUSTAN, and Hartigan; the others, NT-SYS, OSIRIS, SAS, include only the hierarchical procedures. Among biologists there is a clear preference for multilevel techniques in view of their views regarding

evolutionary development; social scientists generally choose single-rank techniques. Although NT-SYS appears best for biologists because of its graphs and manual, the development of criteria for choosing an optimal number of clusters (such as Mojena's Rule One) makes the hierarchical procedures useful.

Linkage Form. Blashfield found that the most popular linkage methods applied were single linkage, complete linkage, average linkage, and minimum variance. Table A-1 lists the options in the programs previously described. Because of its chaining tendency social scientists tend to avoid the single-linkage method; moreover, Monte Carlo tests confirm the poor performance of single linkage in the recovery of compact clusters. Studies that have applied mixture-model tests indicate that average linkage and minimum variance are most robust. Anderberg, CLUSTAN, and HGROUP are the only programs containing Ward's method.

Iterative Partitioning Methods. Seven popular programs perform iterative partitioning cluster analysis: Anderberg, BCTRY, CLUS, CLUSTAN, Hartigan, ISODATA, and MIKCA. The characteristics of these programs are listed briefly in Table A-2. Iterative partitioning methods seek to optimize the homogeneity of clusters. The optimization criteria derive from the well-known matrix identity

$$\mathbf{T} = \mathbf{W} + \mathbf{B}$$

Table A-1. Cluster Methods Offered in Popular Hierarchical Programs.

Program Source	*Single Linkage*	*Complete Linkage*	*Average Linkage*	*Minimum Variance*
Anderberg	+	+	+	+
BMDP (1981)				
P1M[a]	+	+	+	
P2M	+			
CLUSTAN	+	+	+	+
Hartigan	+		+	
HICLUS	+	+		
HGROUP				+
NT-SYS	+	+	+	
OSIRIS (1973)		+		
SAS (1979)	+	+		

[a] Clusters variables (not cases).

Table A-2. Characteristics of Iterative Partitioning Programs.

Program	*Initial Partition*	*Types of Passes*	*Clustering Criteria*	*Number of Clusters*	*Outliers*
Anderberg	Random	*k* means	tr **W**	Fixed	Not permitted
	User-chosen				
BCTRY	User-chosen	*k* means	tr **W**	Variable	Permitted
	Other options				
CLUS	Random	*k* means	tr **W**	Fixed	Not permitted
	User-chosen	Hill climbing	Wilks's lambda		
	Other options	Other	2 others		
CLUSTAN	User-chosen	*k* means	tr **W**	Fixed or variable	Permitted
	Other options	Hill climbing	3 others		
Hartigan		*k* means	tr **W**	Fixed or variable	Not permitted
ISODATA	Other options	*k* means	tr **W**	Variable	Permitted
MIKCA	Random	*k* means	tr **W**	Fixed	Not permitted
	User-chosen	Hill climbing	Wilks's lambda, 2 others		

where **T** is the total dispersion matrix, **W** is the pooled within-groups dispersion matrix, and **B** is the between-groups dispersion matrix.

The first feature that differentiates the partitioning method is the procedure whereby the program is started. Some use a random set; others provide for user-selected seed points; CLUS permits four starting options.

The second feature that distinguishes the various programs is the type of pass used to assign entities to clusters. Anderberg, BCTRY, Hartigan, and ISODATA utilize *k*-means passes only (allocation to the nearest centroid). CLUS, CLUSTAN, and MIKCA also use hill-climbing passes in which an entity is moved from one cluster to the next if the criterion is optimized.

A third characteristic that differentiates the various programs is the statistical clustering criteria employed. The most popular procedure is to minimize the trace of **W** (tr **W**). The trace of a matrix is the sum of its diagonal elements. A second method is to minimize the ratio of $|\mathbf{W}|/|\mathbf{T}|$, which is widely known as Wilks's lambda statistic. This criterion is equivalent to minimizing $|\mathbf{W}|$. A third criterion is known as Roy's largest-latent-root criterion. A fourth criterion is to maximize the trace of $\mathbf{W}^{-1}\,\mathbf{B}$.

A fourth characteristic of the programs is whether the number of clusters must be specified by the user in advance or whether a variable number of clusters is allowed to emerge. CLUSTAN combines clusters specified by the user; BCTRY and ISODATA provide for splitting and merging clusters.

A fifth difference between programs concerns outliers. Outliers are not permitted in Anderberg, CLUS, and Hartigan; other programs assign any entity outside a certain distance of a cluster centroid to a group of outliers that can be added on a later pass. If outliers are permitted, the resulting set of clusters will not be exhaustive. However, there are studies suggesting that outliers should be eliminated at an early stage in order to recover the true structure.

Graphics. The usual form of graphic output from a hierarchical cluster analysis is a tree (dendrogram or phenogram). The tree provides the user with a visual representation of the cluster structure of the hierarchy. The programs and packages differ substantially in the form of their representation, however. Moreover, the output of a tree is not standard for some programs; packages such as NT-SYS

and CLUSTAN require special control cards. According to Aldenderfer and Blashfield (1978), NT-SYS generates the most easily visualized dendrograms from a line printer—at each step it is easy to see which entities join a cluster. The trees of the remaining packages are not so well organized visually. The trees of BMDP1M and P2M are difficult to read since the nodes where entities and clusters merge are less easily observable. Several programs such as HICLUS, OSIRIS, and SAS form a *skyline plot* that is not so obvious as in NT-SYS or Anderberg.

APPENDIX B

Matrix Concepts and Operations

The theory of matrices and determinants originated in the need for solving simultaneous linear equations and for a compact notation for dealing with linear transformations. Matrix algebra is thus an extension of ordinary algebra, and many of the rules of ordinary algebra have analogs in matrix algebra.

A *matrix* is a rectangular table of numbers (or symbols representing numbers) called *elements* or *entries.* The matrix may consist of any number of rows (horizontal arrays) and any number of columns (vertical arrays). If a matrix has only one row it is called a *row vector.* If the matrix consists of a column of elements it is called a *column vector.* A matrix consisting of a single element is an ordinary number or *scalar.*

There are a number of ways in which a table of numbers or symbols can be designated a matrix. Use will be made of brackets as here:

$$\mathbf{A} = \begin{bmatrix} a_{11} & a_{12} & a_{13} \\ a_{21} & a_{22} & a_{23} \\ a_{31} & a_{32} & a_{33} \\ a_{41} & a_{42} & a_{43} \end{bmatrix} = \begin{bmatrix} 1 & 3 & 0 \\ 4 & 2 & 1 \\ 5 & 3 & 4 \\ 1 & -1 & 6 \end{bmatrix}$$

Row and column vectors in this type of *rectangular notation* are as

follows:

$$\mathbf{V} = [v_1 \quad v_2 \quad v_3] \qquad \mathbf{W} = \begin{bmatrix} w_1 \\ w_2 \\ w_3 \end{bmatrix}$$

The *order* of a matrix gives the number of rows and columns (the dimensionality). Matrix **A** is said to be a 4 × 3 table.

Each element can be specified by the row and column of the cell in which the element is located. When symbols are used, the subscripts i and j indicate the position of the element. The first subscript i refers to the row; the second subscript j refers to the column. Thus a_{32} is the element in the third row and second column.

The entire table can be designated in *matrix notation* by a capital letter such as **A**, **B**, **X**, or **Y**. Consider, for example, a set of equations of ordinary algebra:

$$\begin{aligned} x_1 + 2x_2 + x_3 &= y_1 \\ 2x_1 + 3x_2 + 4x_3 &= y_2 \\ 3x_1 + x_2 + 5x_3 &= y_3 \end{aligned}$$

These equations can be expressed in rectangular notation by detaching the coefficients a_{ij} from the variables x_j as follows:

$$\begin{bmatrix} 1 & 2 & 1 \\ 2 & 3 & 4 \\ 3 & 1 & 5 \end{bmatrix} \begin{bmatrix} x_1 \\ x_2 \\ x_3 \end{bmatrix} = \begin{bmatrix} y_1 \\ y_2 \\ y_3 \end{bmatrix}$$

This equation can be summarized in matrix notation as

$$\mathbf{AX} = \mathbf{Y}$$

Matrices can be categorized as data matrices and derived matrices. A data matrix usually consists of N rows for entities and n columns for observations or measurements on n attributes. Data matrices generally have more rows than columns. When the information in data matrices has been transformed by various operations, the tables are said to be derived matrices. Derived matrices are likely to be square or both square and symmetric as in a correlation matrix.

There is an operation in matrix algebra that has no analog in ordinary algebra. From a matrix **A** another matrix can be constructed, the rows of which are columns of **A** and the columns of which are the rows of **A**. The operation is called a *transposition;* the resulting matrix is designated the *transpose* of **A** and is denoted **A**′. For example:

$$\text{If } \mathbf{B} = \begin{bmatrix} 3 & 1 & 0 \\ -2 & 4 & 5 \end{bmatrix} \qquad \text{then } \mathbf{B}' = \begin{bmatrix} 3 & -2 \\ 1 & 4 \\ 0 & 5 \end{bmatrix}$$

Kinds of Matrices

There are various kinds of matrices of importance to social scientists. A *square* matrix has the same number of rows as columns. A square matrix is said to be *symmetric* when $\mathbf{A}' = \mathbf{A}$. Then the elements $a_{ji} = a_{ij}$. Such a matrix is symmetric about the *principal diagonal,* which consists of entries with the same subscripts (a_{11}, a_{22}, . . . , a_{nn}) and runs from the upper left to the lower right corner of the matrix:

$$\begin{bmatrix} 1 & 3 & 4 \\ 3 & 2 & 5 \\ 4 & 5 & 3 \end{bmatrix}$$

A *diagonal* matrix is a special case of a symmetric matrix in which all nondiagonal elements are zero:

$$\mathbf{D} = \begin{bmatrix} a_{11} & 0 & 0 \\ 0 & a_{22} & 0 \\ 0 & 0 & a_{33} \end{bmatrix} \qquad \text{or} \qquad \begin{bmatrix} a_{11} & & \\ & a_{22} & \\ & & a_{33} \end{bmatrix}$$

A *scalar* matrix is a diagonal matrix in which all the diagonal elements are equal. A special case of the scalar matrix is the *identity* matrix in which all the diagonal elements are unity. It is a convention to use the capital letter **I** to designate an identity matrix:

$$\mathbf{I} = \begin{bmatrix} 1 & 0 & 0 \\ 0 & 1 & 0 \\ 0 & 0 & 1 \end{bmatrix}$$

Addition of Matrices

Matrices that are to be added must be of the same order—that is, they must have the same number of rows and columns. Thus the sum of two $m \times n$ matrices is the $m \times n$ matrix each of whose elements is the sum of the corresponding elements of the given matrices:

$$\begin{bmatrix} 3 & 7 \\ 5 & 1 \end{bmatrix} + \begin{bmatrix} 4 & 1 \\ 2 & 6 \end{bmatrix} = \begin{bmatrix} 7 & 8 \\ 7 & 7 \end{bmatrix}$$

The addition of matrices is *commutative* (they may be added in any order):

$$\mathbf{A} + \mathbf{B} = \mathbf{B} + \mathbf{A}$$

Addition is also *associative:*

$$\mathbf{A} + (\mathbf{B} + \mathbf{C}) = (\mathbf{A} + \mathbf{B}) + \mathbf{C}$$

The ordinary quantities of algebra are called *scalars.* The product of a matrix **A** and a scalar k ($k\mathbf{A}$ or $\mathbf{A}k$) is defined as the matrix each of whose elements is k times the elements of **A**. The rule is to multiply each element of the matrix by that scalar.

The rule for subtracting one matrix from another is the same as for addition but represents a special case of the scalar multiplication rule where $k = -1$. Multiply the elements of the matrix **B** by -1 and add the corresponding elements of **A**:

$$\mathbf{A} + (-1)\mathbf{B} = \mathbf{C}$$

Multiplication of Matrices

In order to multiply two matrices and form a product **AB**, the number of columns in **A** must equal the number of rows in **B**. This rule is necessary because there must be an entry in each row of matrix **A** to match with each entry in a column of matrix **B**.

The rule for matrix multiplication can be stated concisely: Multiply row by column. Given two matrices **A** and **B**, to find the

entry in the ith row and jth column of the *product matrix* **AB** (called **C**) multiply each entry in the ith row of the left-hand factor **A** (called the *premultiplier*) by the corresponding entry in the jth column of the right-hand factor **B** (called the *postmultiplier*). Then add all the resulting terms. To illustrate:

$$\mathbf{AB} = \begin{bmatrix} 1 & 3 & 2 \\ 2 & 4 & 0 \end{bmatrix} \begin{bmatrix} 1 & 5 \\ 2 & 4 \\ 3 & 2 \end{bmatrix} = \begin{bmatrix} 13 & 21 \\ 10 & 26 \end{bmatrix} = \mathbf{C}$$

The procedure for the product **AB** is as follows:

1. Multiply each element of the first row of **A** by the corresponding element of the first column of **B** and add the products:

$$\mathbf{C}_{11} = 1(1) + 3(2) + 2(3) = 13$$

2. Multiply each element of the first row of **A** by the corresponding element of the second column of **B** and add the products:

$$\mathbf{C}_{12} = 1(5) + 3(4) + 2(2) = 21$$

3. Similarly, cross-multiply the elements of the second row of **A** by the elements of the first column of **B** and add:

$$\mathbf{C}_{21} = 2(1) + 4(2) + 0(3) = 10$$

4. Cross-multiply the second row of **A** by the second column of **B** to get the product:

$$\mathbf{C}_{22} = 2(5) + 4(4) + 0(2) = 26$$

The multiplication of ordinary numbers is *commutative;* that is to say, $ab = bc$. The multiplication of matrices is *noncommutative;* in general, $\mathbf{AB} \neq \mathbf{BA}$. To reduce ambiguity in referring to multiplication of **A** and **B**, it is useful to say either "**A** is postmultiplied by **B**" or "**B** is premultiplied by **A**."

In multiplication of numbers, the triple product of a, b, and c is $(ab)c = a(bc)$. This property of *associativity* is also true for

matrices. Given three matrices **A**, **B**, and **C**, the triple product (**AB**)**C** = **A**(**BC**). Another property common to multiplication of matrices and multiplication of numbers is *distributivity* over addition:

$$\mathbf{A}(\mathbf{B} + \mathbf{C}) = \mathbf{AB} + \mathbf{AC}$$

and

$$(\mathbf{B} + \mathbf{C})\mathbf{A} = \mathbf{BA} + \mathbf{CA}$$

A special case of matrix multiplication concerns premultiplication of a column vector **V** by a row vector **W'**:

$$\mathbf{W'V} = [2 \quad 4]\begin{bmatrix} 3 \\ 1 \end{bmatrix} = 2(3) + 4(1) = 10$$

The product **W'V** is called the *scalar product* because the product is a scalar quantity. This product is also referred to as a *minor product*, an *inner product*, and *dot product* of the two vectors. A minor product is simply the sum of the products of corresponding elements of the two vectors.

A *major product* is a column vector times a row vector:

$$\begin{bmatrix} 2 \\ 4 \end{bmatrix}[3 \quad 1] = \begin{bmatrix} 2 \times 3 & 2 \times 1 \\ 4 \times 3 & 4 \times 1 \end{bmatrix} = \begin{bmatrix} 6 & 2 \\ 12 & 4 \end{bmatrix}$$

Whereas the minor product is always a scalar quantity, the major product of two vectors is never a scalar quantity—it is a matrix.

Inverse of a Matrix

A natural question to ask of matrices is whether there are analogs to division and the reciprocal of a matrix (a multiplicative inverse). In ordinary arithmetic, for example, the reciprocal of 5 is $\frac{1}{5}$ and $5 \times \frac{1}{5} = 1$. Thus for any number $x \neq 0$, there exists a reciprocal such that $x(x^{-1}) = 1$. With certain restrictions, matrices do have inverses or reciprocals. Our discussion will be confined to square matrices because nonsquare matrices require advanced procedures for solution. An inverse of a matrix **A**, if it exists, is denoted $\mathbf{A}^{-1}$. For

square matrices, premultiplying or postmultiplying by an inverse yields an identity matrix:

$$\mathbf{A}^{-1}\mathbf{A} = \mathbf{A}\mathbf{A}^{-1} = \mathbf{I}$$

In order to compute an inverse, knowledge is needed of *determinants.* The following paragraphs explain the basic operations involved in computing an inverse; ordinarily a computer is used.

The Concept of Determinant. Every square matrix has a determinant that is denoted by a square table bounded by two straight lines—or, more briefly, denoted by $|\mathbf{A}|$. It represents the sum of $n!$ terms, each of which is a product of n elements with only one from each row and one from each column. Here the symbol $n!$ (n factorial) means the product of a series of numbers 1, 2, 3, . . . , n. For example, the second-order determinant

$$|\mathbf{A}| = \begin{vmatrix} a_{11} & a_{12} \\ a_{21} & a_{22} \end{vmatrix}$$

is the sum of the products $a_{11}a_{22} - a_{21}a_{12}$.

The calculation of a determinant is best explained by example. Consider a third-order determinant

$$|\mathbf{A}| = \begin{vmatrix} a_{11} & a_{12} & a_{13} \\ a_{21} & a_{22} & a_{23} \\ a_{31} & a_{32} & a_{33} \end{vmatrix}$$

A determinant of the nth order consists of n^2 elements, each identified by two subscripts. The first subscript represents the number of the row; the second represents the number of the column.

To evaluate the determinant, form all possible products of n elements such that each term includes only one entry from each row and one from each column. If the matrix is of order 3×3, for example, there are 3! or six terms. To determine the sign of a term, arrange the n elements of each term in a natural ascending order according to row subscripts. Then count the number of inversions or transpositions involving column subscripts. An inversion has occurred when a larger subscript precedes a smaller one. In the second term below, the column subscripts are 2-3-1 and two inversions

occur—subscript 2 comes before 1 and subscript 3 before 1. Then multiply each product that has an odd number of inversions by -1. If no inversions or an even number of inversions are involved, multiply by $+1$ or leave the product as is. The six terms for the 3×3 determinant are as follows:

$$\begin{aligned} &+a_{11}a_{22}a_{33} + a_{12}a_{23}a_{31} + a_{13}a_{21}a_{32} \\ &-a_{13}a_{22}a_{31} - a_{11}a_{23}a_{32} - a_{12}a_{21}a_{33} = |\mathbf{A}| \end{aligned}$$

The sum of these $n!$ terms is the value of the determinant $|\mathbf{A}|$.

Expansion of Determinants by Cofactors. This process of finding the numerical value of **A** is tedious and likely to involve error. A more rapid procedure is expansion by cofactors. To compute by this process, definitions of minors and cofactors are needed. A *minor* of an element (a_{ij}) of the square matrix **A** is the determinant of a submatrix obtained by deleting the ith row and the jth column of **A**. For example, the minor of the entry a_{21} in third-order matrix is

$$M_{11} = \begin{vmatrix} a_{12} & a_{13} \\ a_{32} & a_{33} \end{vmatrix}$$

The *cofactor* of an element a_{ij} of a square matrix **A** is the product of the minor of a_{ij} and $(-1)^{i+j}$. The cofactor is called a signed minor and is often denoted by A_{ij}. The cofactor of element a_{21} is

$$A_{21} = (-1)^{2+1} \begin{vmatrix} a_{12} & a_{13} \\ a_{32} & a_{33} \end{vmatrix} = -(a_{12}a_{33} - a_{32}a_{13})$$

It is often convenient to assign a plus sign and a minus sign to alternate cells in a determinant. The rule is to designate the upper left cell as positive and other cells as alternately negative and positive. The sign arrangement of a 3×3 determinant is

$$\begin{vmatrix} + & - & + \\ - & + & - \\ + & - & + \end{vmatrix}$$

The sign of a cell is called its *position sign.* If the element is a_{ij}, then the position sign is $(-1)^{i+j}\ a_{ij}$. When the exponent is odd, the sign is negative; when the exponent is even, the sign is positive.

By this method of expansion the value of $|\mathbf{A}|$ can be determined by multiplying *any* array (row or column) of elements by their cofactors and adding the terms. For example, the value of $|\mathbf{A}|$ may be obtained by expanding by its first column:

$$a_{11}\begin{vmatrix} a_{22} & a_{23} \\ a_{32} & a_{33} \end{vmatrix} - a_{21}\begin{vmatrix} a_{12} & a_{13} \\ a_{32} & a_{33} \end{vmatrix} + a_{31}\begin{vmatrix} a_{12} & a_{13} \\ a_{22} & a_{23} \end{vmatrix}$$
$$= a_{11}[a_{22}a_{33} - a_{32}a_{23}] - a_{21}[a_{12}a_{33} - a_{32}a_{13}] + a_{31}[a_{12}a_{23} - a_{22}a_{13}]$$

Computing an Inverse. To compute an *inverse* of the matrix **A**, its determinant must be unequal to zero ($|\mathbf{A}| \neq 0$). The matrix is then said to be *nonsingular.* If $\mathbf{A} = 0$, the matrix is *singular* and there is no inverse. The steps are as follows:

1. Solve for the value of the determinant by multiplying each element in one array by its cofactors and summing the terms.
2. Write the matrix $\mathbf{M}_{ij}$ with elements m_{ij}, the first minors of **A**.
3. Write the matrix $\mathbf{A}_{ij}$ of the cofactors of the elements of **A.** These are determined by reversing signs of alternate elements of **M**. Here $\mathbf{A}_{ij} = (-1)^{i+j}\mathbf{M}_{ij}$.
4. Construct the *adjoint* of **A** or $\text{adj}(\mathbf{A}) = \mathbf{A}'_{ij}$ by writing the transpose of $\mathbf{A}_{ij}$.
5. Divide each element of the adjoint by the value of the determinant. This matrix is the inverse $\mathbf{A}^{-1}$.
6. Check the accuracy of the operation by multiplying **A** by $\mathbf{A}^{-1}$:

$$\mathbf{A}\mathbf{A}^{-1} = \mathbf{I}$$

An Illustration. Suppose we expand the determinant around the first column of the matrix:

$$\mathbf{A} = \begin{bmatrix} 1 & -1 & 3 \\ 2 & 2 & 1 \\ 2 & 1 & 4 \end{bmatrix}$$

The entries of the first column of **A** are 1, 2, and 2:

$$|\mathbf{A}| = 1\begin{vmatrix} 2 & 1 \\ 1 & 4 \end{vmatrix} - 2\begin{vmatrix} -1 & 3 \\ 1 & 4 \end{vmatrix} + 2\begin{vmatrix} -1 & 3 \\ 2 & 1 \end{vmatrix}$$
$$= 1(8 - 1) - 2(-4 - 3) + 2(-1 - 6)$$
$$= 7$$

The matrix of minors **M** is

$$\mathbf{M}_{ij} = \begin{bmatrix} 7 & 6 & -2 \\ -7 & -2 & 3 \\ -7 & -5 & 4 \end{bmatrix}$$

The matrix of signed minors of **M** called cofactors is

$$\mathbf{A}_{ij} = \begin{bmatrix} 7 & -6 & -2 \\ 7 & -2 & -3 \\ -7 & 5 & 4 \end{bmatrix}$$

The transpose of $\mathbf{A}_{ij}$ is the adj $(\mathbf{A}_{ij})$:

$$\text{adj}(\mathbf{A}_{ij}) = \begin{bmatrix} 7 & 7 & -7 \\ -6 & -2 & 5 \\ -2 & -3 & 4 \end{bmatrix}$$

The inverse of the matrix **A** is then

$$\mathbf{A}^{-1} = \begin{bmatrix} 7/7 & 7/7 & -7/7 \\ -6/7 & -2/7 & 5/7 \\ -2/7 & -3/7 & 4/7 \end{bmatrix}$$

To illustrate the process in matrix algebra, try solving for the unknowns **X** in $\mathbf{AX} = \mathbf{B}$. Premultiply each side of $\mathbf{AX} = \mathbf{B}$ by $\mathbf{A}^{-1}$:

$$\mathbf{A}^{-1}(\mathbf{AX}) = \mathbf{A}^{-1}\mathbf{B}$$

But $\mathbf{A}^{-1}\mathbf{A} = \mathbf{I}$, so

$$\mathbf{IX} = \mathbf{A}^{-1}\mathbf{B}$$
$$\mathbf{X} = \mathbf{A}^{-1}\mathbf{B}$$

Theorems on Transpose and Inverse of Products of Matrices. There are two important theorems in matrix operations. First, the transpose of a product of matrices is equal to their transposes taken in reverse order:

$$(\mathbf{ABC})' = \mathbf{C}'\mathbf{B}'\mathbf{A}$$

Second, the inverse of a product of matrices is the product of their inverses taken in reverse order:

$$(\mathbf{ABC})^{-1} = \mathbf{C}^{-1}\mathbf{B}^{-1}\mathbf{A}^{-1}$$

An Algebraic Illustration

The economy of matrix operations can be illustrated in the process of computing sums of squares and sums of cross-products as well as correlations. Suppose we are given an $N \times 3$ table of mean-corrected deviation scores of N persons on three attributes:

$$\mathbf{x} = \begin{bmatrix} x_{11} & x_{12} & x_{13} \\ x_{21} & x_{22} & x_{23} \\ \cdot & \cdot & \cdot \\ \cdot & \cdot & \cdot \\ \cdot & \cdot & \cdot \\ x_{N1} & x_{N2} & x_{N3} \end{bmatrix}$$

where in x_{ij}, $i = 1, 2, \ldots, N$ and $j = 1, 2, 3$. Now if $\mathbf{x}$ is multiplied by its transpose $\mathbf{x}'$, a matrix of sums of squares and sums of cross-products will result:

$$\mathbf{xx}' = \begin{bmatrix} \sum x_1^2 & \sum x_{1i}x_{2i} & \sum x_{1i}x_{3i} \\ \sum x_{2i}x_{1i} & \sum x_2^2 & \sum x_{2i}x_{3i} \\ \sum x_{31}x_{1i} & \sum x_{3i}x_{2i} & \sum x_3^2 \end{bmatrix} = \mathbf{CP}$$

A matrix of covariances will result if the matrix **CP** is multiplied by $1/N$, the number of cases. Thus

$$\frac{\mathbf{CP}}{N} = \begin{bmatrix} s_1^2 & r_{12}s_1s_2 & r_{13}s_1s_3 \\ r_{21}s_2s_1 & s_2^2 & r_{23}s_2s_3 \\ r_{31}s_3s_1 & r_{31}s_3s_2 & s_3^2 \end{bmatrix} = \mathbf{C}$$

If the matrix of covariance is premultiplied and postmultiplied by an inverse ($\mathbf{D}^{-1}$) of a diagonal matrix **D** whose diagonal elements contain the standard deviations of the attributes, this procedure

$$\mathbf{D}^{-1}\mathbf{C}\mathbf{D}^{-1}$$

yields

$$\mathbf{R} = \begin{bmatrix} 1.0 & r_{12} & r_{13} \\ r_{21} & 1.0 & r_{23} \\ r_{31} & r_{32} & 1.0 \end{bmatrix}$$

If we begin with a matrix of standard scores **Z**, then the cross-product matrix becomes

$$\mathbf{R} = \frac{1}{N} \mathbf{ZZ}'$$

Glossary

Actuarial prediction. From contingent frequency tables of predictor attributes and qualities, actuarial procedures derive estimates of probability that entities are correctly classified.

Additive clustering. A method of representation of similarities between objects as a simple additive function of discrete overlapping properties. The method allows for overlapping clusters in which an object can belong to more than one cluster.

Agglomerative clustering. A clustering procedure in which entities are successively merged into more inclusive groups.

Attribute. A property capable of further division; a quantitative variable or unidimensional continuum. (See *Quality.*)

A-space. Attribute space formally has n dimensions, one for each attribute; N entity points are located in this space.

Average-linkage hierarchical clustering. An agglomerative procedure in which the clusters merged at each stage are the two whose members have the least average dissimilarity. The distance between clusters is defined as the average distance between all pairs of entities in the two clusters.

Basic structure matrix. Any data matrix may be decomposed into its basic structure, which is a triple of three matrices $\mathbf{P} \, \mathbf{\Lambda} \, \mathbf{Q}'$. Matrices $\mathbf{P}$ and $\mathbf{Q}$ are orthonormal by columns; $\mathbf{\Lambda}$ is diagonal with K positive diagonal entries.

Block model. A binary data matrix arranged to reveal blocks of high and low density values (1's and 0's) for a set of attributes. By means of the CONCOR algorithm the rows and columns of the data matrix are permuted and partitioned into homogeneous subsets. See *Intercolumnar correlation analysis.*

Centering. A process consisting of subtracting (1) the variable means from each variable score or (2) the entity profile means from profile scores. The scores that result are called deviation scores.

Centroid hierarchical clustering. An agglomerative procedure by which clusters are merged according to the distance between their centroids, the clusters with the smallest distance being merged first. The distance between clusters is defined as the distance between the cluster centroids.

Chained or connected cluster. A subset of entities linked by asymmetric transitive order relationships. In attribute space the cluster is long, straggly, serpentine, or amoeboid.

Classification. A systematic division or arrangement into categories or groups.

Clustering. The grouping of entities into homogeneous subsets on the basis of similarity across a sample of attributes.

Common factor. A hypothetical underlying variable common to two or more observed variables.

Communality. The variance of an observed variable that represents the common factors. In an orthogonal factor model it is equal to the sum of its squared factor coefficients.

Compact cluster. A subset of entities characterized by the properties of isolation and coherence (mutual similarity). In attribute space the cluster is circular, convex, and separated from other clusters.

Complete-linkage hierarchical clustering. An agglomerative procedure in which the clusters merged at each stage are the two with the largest intracluster distance. The distance between clusters is defined as the distance between their most remote pair of entities.

Configural scale. The scale is conceived of as a polynomial function of t dichotomous variables, each weighted by a regression coefficient. The function is used to predict a quantitative criterion C'. Included are cross-product terms to represent possible interactions.

Congruency coefficient. The normalized cross-product of two sets of raw scores summed across k variables.

Conjoint set. A set comprised of all entities combined. A disjoint set is the set of all entities considered singly; no two are combined.

Continuous variable. A variable with an uncountable range and an infinite number of closely spaced inbetween values. Examples are age, time, and test score.

Convex cluster. A set of points in n-dimensional attribute space such that for every pair of points in the set the line segment joining them is also in the set.

Correlation. A measure of the degree of association between two variables. The term usually refers to the product-moment correlation, but it is also used to describe any association between variables.

Covariance. A measure of the degree to which two variables covary. It is measured as the sum of cross-products of the scores expressed as deviations from their respective means divided by the number of cases.

Criterion. What we wish to predict. It is the dependent variable that is to be predicted from a set of independent predictors.

Density search technique. Clustering methods that search for regions of high density or modes in attribute space.

Discrete variable. A variable with a finite or countable range that takes only point values such as 0, 1, 2, and so on. A binary variable is a special case; it has only two values.

Discriminant function analysis. A statistical procedure for classifying N entities into g mutually exclusive known categories on the basis of their variable scores. The goal is to minimize the number of misclassifications.

Disjoint set. See *Conjoint set.*

Distance measure. A measure of profile dissimilarity defined as the square root of the sum of squared differences between two entities across k variables.

Divisive clustering. A clustering procedure in which a set of N entities is successively partitioned into finer and finer subsets.

Dominance relation. An ordinal, transitive relation between two entities—A beats B in tennis, for example, or tone C is louder than D.

Dummy variable. A binary coded vector in which members of a category are coded *1* while all others are coded *0*. The number of dummy variables for k categories is $k - 1$. Examples are coding for religion, marital status, or sex.

Eigenvalue (latent or characteristic root). A measure of variance accounted for by a given principal component or principal axis. The value represents the length of the eigenvector.

Eigenvector (latent vector). The vector associated with each eigenvalue. When normalized these vectors become principal components. The vector's coordinates specify its direction.

Elevation. Mean score of an entity over all measures in the score profile.

Entity. A person or object of ordinary experience that possesses quantitative or qualitative properties. Entity refers to the elements of the set being classified (case, data unit, object).

Entity space. Entity space has formally N dimensions, one for each entity. In E-space n points are located to represent the n attributes.

Factor. A dimension of individual differences; a scientific construct intended to explain what is common to a set of attributes.

Factor analysis. A term that refers to several mathematical proce-

dures for analyzing the relationships among a set of variables and explaining them in terms of a reduced set of hypothetical unmeasured variables.

Factor loading. A popular term that refers to the coefficients of the common factors in a factor pattern matrix or the correlations of each variable with each factor given in a factor structure. When the factors are uncorrelated, the two definitions are identical.

Factor score. An estimate of an underlying factor formed from a weighted linear combination of observed variables.

Gramian matrix. A square symmetric matrix with positive weights in the diagonal cells.

Hierarchical or multilevel techniques. *Agglomerative* techniques begin with a disjoint set and merge entities or clusters into successively fewer clusters until all are combined. *Divisive* techniques begin with a complete set and subdivide the conjoint set into two subsets and each subset successively into smaller and smaller subsets.

Hierarchical scheme. A nested family of clusters that can be represented as a tree beginning at the top branches and merging successively until the trunk is reached.

Identification. The allocation of unidentified entities to established categories.

Intercolumnar correlation analysis. A divisive clustering procedure that starts with a matrix of interassociation between *N* entities. The correlations between columns are computed and the process is iterated until nearly all coefficients are $+1$ or -1. The final matrix is subdivided into submatrices of $+1$ and -1. The method is also called CONCOR (Convergence of Iterated Correlations).

Interval scale. A scale characterized by a common and constant unit of measurement. The zero point and the unit of measurement are arbitrary, however.

Intraclass correlation. A measure of association based on the ratio of mean squares between variables minus the mean squares within variables over the total mean squares.

Ipsative. A method of assessment that yields estimates dependent on the entity's own scores on other variables but independent of the scores of other entities. The sum of scores over attributes for each entity is a constant.

k-means cluster analysis. A sample of N is sorted or partitioned into k clusters on the basis of the shortest distance between the entity and the k cluster means.

k-space. Attribute space defined by a reduced set of k relatively independent dimensions. See *A-space.*

Least-squares solution. A solution that minimizes the squared deviations between the observed values and the predicted values.

Linear regression. A procedure for finding a set of optimum weights for a set of variables to predict some dependent variable (criterion). The least squares regression line is found by minimizing the errors of estimate.

Link. A pair correlation at or above an arbitrary minimum or the greatest correlation an entity has with other entities. For pair distances, a link is a specified maximum distance or the shortest distance between an entity and other entities.

Matching coefficients. Measures of association that assess agreement between two entities over n binary variables. Presence or absence of a characteristic is coded 1 or 0. Coefficients take values of 0 to 1.

Metric. A system of units of measurement.

Minimum-variance method. Ward's hierarchical clustering procedure in which groups are formed such that the sum of squared within-group deviations about the group mean is minimized for all variables.

Minkowski metrics. A general class of distance functions based on one parameter. Euclidean and city-block distances are special cases.

Mixture problem. Given a data set consisting of a mixture of samples from several populations, the task is to resolve the mixture and identify the unknown number of populations and their characteristics.

Monothetic. A divisive technique based on possession of a single specified binary attribute that is used to bisect the set of entities at each stage. (See *Polythetic.*)

Monotone transformation. Any transformation of scale values that preserves their rank order.

Monotonic relation. A relationship between variables which uniformly increases or decreases. Examples are the normal ogive and exponential curves.

Monte Carlo study. A procedure whereby various sample properties based on complex statistical models are simulated.

Multilevel hierarchical techniques. See *Hierarchical.*

Multiple correlation coefficient. The maximum correlation to be expected between the dependent variable and a linear additive combination of independent variables.

Multiple regression analysis. A system of data analysis used whenever a quantitative variable is to be studied as a function of a set of independent variables (quantitative or qualitative).

Natural classes. Groupings that correspond closely to the underlying structure. The term also refers to classes that possess broad scientific import and generality.

Nested relationships. If two sets have an identity relationship, each includes the other (mutually nested). If two sets have an inclusion relationship, all entities in one set are included in the other but not vice versa (one is nested in the other).

Nominal scale. A system of numbers or symbols used to identify or categorize a set of entities (objects, persons, stimuli)—for example, religion, ethnic group, marital status.

Nonmonotonic relationship. A relationship between variables that changes direction and doubles back on itself.

Normalization. The process in which a vector is converted to unit length. The values are divided by the square root of their sum of squares.

Normative. A method of assessment that yields units such that an entity's scores are statistically dependent on the score of other entities in the sample and independent of other scores of the entity. Standard scores and percentiles are examples.

Ordinal scale. A scale in which the relation "greater than" holds for all pairs of classes so that a complete rank order arises—for example, rating the hardness of stones. The relation is irreflexive, asymmetric, and transitive.

Ordination. Procedures, such as principal-component analysis, that permit the mapping of N entities in low-dimensional attribute space. The points can then be visually inspected to identify clusters.

Orthogonal factors. Orthogonal factors (vectors) lie at right angles to each other; oblique factors lie at acute or obtuse angles to each other. Orthogonal factors are uncorrelated whereas oblique factors are correlated.

Outlier. Any case in a relationship that is not part of the greater majority of cases; it lies well below, above, or outside the relationship.

Partition. A family of mutually exclusive clusters such that each entity lies in just one member of the partition.

Partitioning. The process of separating a set of N entities into g mutually exclusive clusters. In optimal partitioning the procedure seeks to maximize some criterion function.

Phenetic relationship. Similarity based on a set of phenotypic characteristics of the entities under study.

Polythetic. A divisive method of grouping based on several attributes. (See *Monothetic.*)

Principal-axis factoring. A method of analyzing a matrix of correlation with adjusted diagonal values. The aim is to decompose the correlations into common factors.

Principal components. Linear combinations of observed variables such that each is orthogonal to every other one and each accounts for the maximum amount of variance in the matrix.

Profile. The vector of values (scores) that characterize an entity in terms of k descriptive variables. Profile vectors differ in mean (elevation), scatter (length), and orientation (or shape) in k-dimensional space.

Promax. A method of oblique rotation in which an orthogonal rotational solution (for example, varimax) is used to obtain a least-squares fit to an ideal oblique solution.

Proximity relation. A measure of nearness or similarity between two entities. The distance relation is symmetric.

Q analysis. A dimensional analysis of the $N \times N$ matrix **Q** of correlations among entities across n attributes. The analysis isolates r entity factors.

Q sort. A procedure in which a set of items representing some universe of attributes is sorted into a number of categories (usually 9 or 11). The items are sorted in order of salience, significance, or representativeness for the entity from least to most characteristic.

Quality. A property not capable of further division (such as red or triangular).

R analysis. A dimensional analysis of the $n \times n$ matrix **R** of correlations among attributes across N entities. The analysis yields the coordinates of the N attributes on r coordinate axes.

Rank of a matrix. The number of linearly independent columns or rows of a matrix.

Ratio scale. An interval scale with a true zero point as its origin. Multiplication by a constant is the only permissible transformation.

Raw score. An absolute measure that is statistically independent of other scores of the entity and independent of the scores of other individuals (for example, a person's weight). Such scores are not referred to any normative group.

Scale types. Scales are characterized in terms of their admissible transformations: nominal, ordinal, interval, and ratio.

Scaling. The process of assigning numbers to entities or properties.

Scatter. A measure of score dispersion in a profile defined as the square root of the sum of squared deviation scores about the entity's own mean.

Scree test. A rule of thumb for determining the number of significant components to retain. It is based on the plot of eigenvalues against the components extracted.

Set. A collection of entities considered together as a whole.

Shape. The rank order of scores within a profile. In geometric terms, shape refers to the orientation or direction of a score vector in n-dimensional space.

Simple structure. A factor structure with certain properties: Each variable is defined by only a few of the common factors, and each common factor is defined by only some of the variables.

Simultaneous partitioning. Techniques of clustering that sort data into multiple clusters and use iteration and relocation to satisfy an optimum criterion.

Single-level successive clustering. Methods that successively form nonhierarchical clusters without iteration or relocation.

Single-linkage hierarchical clustering. An agglomerative procedure in which at each stage the two clusters are merged that have the least distance between their closest members.

Stopping rule. A criterion for determining a stopping point in hierarchical grouping. This point in the process yields a level with an optimum number of clusters present.

Taxon (plural: taxa). A taxonomic group of any rank; a set of objects related by an equivalence relation.

Taxonomic hierarchy. A nested sequence of partitions in which each partition is assigned a rank. The elements of the partition are called taxa (for example, species, genus, family, order, class).

Tree (dendrogram or phenogram). A two-dimensional diagram that shows the fusions or partitions made at successive levels in the agglomerative and divisive methods of clustering.

Triangular inequality. An axiom that requires that the sum of the distances between two points (X and Z) in a triangle and a third point (Y) is always greater than or equal to the distance between these two points.

Type. A cluster distinguished by a common set of characteristics; a genus or species that exemplifies the essential characteristics of a group.

Ultrametric inequality. The ultrametric inequality requires that triples of distances satisfy $D(x,z) \leq \max[D(x,y),D(z,y)]$. Such triples of distances must form an equilateral or isosceles triangle where the base is shorter than the two equal sides.

Variance. A measure of the dispersion of scores on a variable defined as the sum of squared deviations from the mean divided by the number of cases.

Variates. Random variables that take on values over some continuous region are called continuous random variables or continuous variates. Both terms are used synonymously.

Varimax. A method of orthogonal rotation that simplifies the factor structure by maximizing the variance of the squared loadings of each of the columns of the factor pattern matrix.

Vector. A quantity having both magnitude (length) and direction. It may be represented as an arrow and defined by a set of coordinates in attribute space or entity space.

References

Adanson, M. *Familles des Plants.* Vol. 1. Paris, 1763.

Aldenderfer, M. S., and Blashfield, R. K. "Computer Programs for Performing Hierarchical Cluster Analysis." *Applied Psychological Measurement,* 1978, *2,* 403–411.

American Psychiatric Association. *Diagnostic and Statistical Manual of Mental Disorders.* (3rd ed.) DSM-III. Washington, D.C.: American Psychiatric Association, 1980.

Anderberg, M. R. *Cluster Analysis for Applications.* New York: Academic Press, 1973.

Anderson, N. H. "Scales and Statistics: Parametric and Nonparametric." *Psychological Bulletin,* 1961, *58,* 305–516.

Arabie, P. "Clustering Representations of Group Overlap." *Journal of Mathematical Sociology,* 1977, *5,* 113–128.

Arabie, P., and Boorman, S. A. "Blockmodels: Developments and Prospects." In Herschel C. Hudson and associates (Eds.), *Classifying Social Data: New Applications of Analytic Methods for Social Science Research.* San Francisco: Jossey-Bass, 1982.

Arabie, P., and Carroll, J. D. "MAPCLUS: A Mathematical Programming Approach to Fitting the ADCLUS Model." *Psychometrika,* 1980, *45,* 211–235.

Arabie, P., Boorman, S. A., and Levitt, P. R. "Constructing Block-

models: How and Why." *Journal of Mathematical Psychology,* 1978, *17,* 21–63.

Arabie, P., and others. "Overlapping Clustering: A New Method for Product Positioning." *Journal of Marketing Research,* 1981, *18,* 310–317.

Bailey, K. D. "Cluster Analysis." In David R. Heise (Ed.), *Sociological Methodology 1975.* San Francisco: Jossey-Bass, 1974.

Baker, F. B., and Hubert, L. J. "Measuring the Power of Hierarchical Cluster Analysis." *Journal of the American Statistical Association,* 1975, *70,* 31–38.

Bales, R. F., Cohen, S. P., and Williamson, S. A. *SYMLOG: A System for the Multiple Level Observation of Groups.* New York: Free Press, 1979.

Ball, G. H. *Classification Analysis.* Menlo Park, Calif.: Stanford Research Institute, 1970.

Ball, G. H., and Hall, D. J. *ISODATA: A Novel Method of Data Analysis and Pattern Classification.* Menlo Park, Calif.: Stanford Research Institute, 1965.

Bar, A. J., Goodnight, J. H., and Sall, J. P. *SAS User's Guide.* Raleigh: N.C.: SAS Institute, 1979.

Bartko, J. J. "On Various Intraclass Correlation Reliability Coefficients." *Psychological Bulletin,* 1976, *83,* 762–765.

Bartko, J. J., Strauss, J. S., and Carpenter, W. T. "An Evaluation of Taxometric Techniques for Psychiatric Data." *Classification Society Bulletin,* 1971, *2,* 2–28.

Bayne, C. K., and others. "Monte Carlo Comparisons of Selected Clustering Procedures." *Pattern Recognition,* 1980, *12,* 51–62.

Beebe-Center, J. G. *Pleasantness and Unpleasantness.* New York: D. Van Nostrand, 1932.

Blalock, H. M., Jr. *Social Statistics.* New York: McGraw-Hill, 1979.

Blashfield, R. K. "Mixture Model Tests of Cluster Analysis: Accuracy of Four Hierarchical Agglomerative Methods." *Psychological Bulletin,* 1976a, *83,* 377–388.

Blashfield, R. K. "Questionnaire on Cluster Analysis Software." *Classification Society Bulletin,* 1976b, *3,* 25–42.

Blashfield, R. K. "The Growth of Cluster Analysis: Tryon, Ward, and Johnson." *Multivariate Behavioral Research,* 1980, *15,* 439–458.

Blashfield, R. K. "Reply to Lorr and Reanalysis." *Applied Psychological Measurement,* 1981, *5,* 75–76.

Blashfield, R. K., and Aldenderfer, M. S. "The Literature on Cluster Analysis." *Multivariate Behavioral Research,* 1978, *13,* 271–295.

Blashfield, R. K., and Morey, L. C. "A Comparison of Four Clustering Methods Using MMPI Monte Carlo Data." *Applied Psychological Measurement,* 1980, *4,* 57–64.

Boyce, A. J. "Mapping Diversity." In A. J. Cole (Ed.), *Numerical Taxonomy.* New York: Academic Press, 1969.

Breiger, R. L., Boorman, S. A., and Arabie, P. "An Algorithm for Clustering Relational Data, with Application to Social Network Analysis and Comparison with Multidimensional Scaling." *Journal of Mathematical Psychology,* 1975, *12,* 328–383.

Burke, D. J. "Additive Scales and Statistics." *Psychological Review,* 1953, *60,* 73–75.

Burket, G. R. "A Study of Reduced Rank Models for Multiple Prediction." *Psychometric Monograph* 12. Richmond, Va.: William Byrd Press, 1964.

Burt, C. "Correlations Between Persons." *British Journal of Psychology,* 1937, *28,* 167–185.

Burt, C. *The Factors of the Mind.* London: London University Press, 1940.

Burt, C., and Stephenson, W. "Alternative Views on Correlations Between Persons." *Psychometrika,* 1939, *4,* 269–281.

Carmichael, J. W., and Sneath, P.H.A. "Taxometric Maps." *Systematic Zoology,* 1969, *18,* 402–415.

Carroll, J. D. "Individual Differences and Multidimensional Scaling." In Roger N. Sheperd, A. K. Romney, and S. Nerlove (Eds.), *Multidimensional Scaling.* Vol. 1. New York: Seminar Press, 1972.

Carroll, J. D. "Spatial, Non-spatial, and Hybrid Models for Scaling." *Psychometrika,* 1976, *41,* 439–463.

Carroll, J. D., and Chang, J. J. "Analysis of Individual Differences in Multidimensional Scaling via *N*-Way Generalizations of Eckart–Young Decomposition." *Psychometrika,* 1970, *35,* 283–319.

Cattell, R. B. "A Note on Correlation Clusters and Cluster Search Methods." *Psychometrika,* 1944, *9,* 169–184.

Cattell, R. B. "r_p and Other Coefficients of Pattern Similarity." *Psychometrika,* 1949, *14,* 279–298.

Cattell, R. B. "The Scree Test for the Number of Factors." *Multivariate Behavioral Research,* 1966, *1,* 245–276.

Cattell, R. B. *The Scientific Use of Factor Analysis.* New York: Plenum Press, 1978.

Cattell, R. B., and Coulter, M. A. "Principles of Behavioral Taxonomy and the Mathematical Basis of the Taxonomy Computer Program." *British Journal of Mathematical and Statistical Psychology,* 1966, *19,* 237–269.

Clifford, D.H.T., and Stephenson, W. *An Introduction to Numerical Classification.* New York: Academic Press, 1975.

Cohen, J. "A Coefficient of Agreement for Nominal Scales." *Educational and Psychological Measurement,* 1960, *20,* 37–46.

Cohen, J. "r_c: A Profile Similarity Coefficient Invariant over Variable Reflection." *Psychological Bulletin,* 1969, *71,* 281–284.

Cohen, J., and Cohen, P. *Applied Multiple Regression/Correlation Analysis for the Behavioral Sciences.* New York: Lawrence Erlbaum, 1975.

Coombs, C. H., and Satter, G. A. "A Factorial Approach to Job Families." *Psychometrika,* 1949, *14,* 33–42.

Cormack, R. M. "A Review of Classification." *Journal of the Royal Statistical Society* (Series A), 1971, *134,* 321–367.

Cronbach, L. J., and Gleser, G. C. "Assessing Similarity Between Profiles." *Psychological Bulletin,* 1953, *50,* 456–473.

Cunningham, K. M., and Ogilvie, J. C. "Evaluation of Hierarchical Grouping Techniques: A Preliminary Study." *Computer Journal,* 1972, *15,* 209–213.

Dawes, R. M. "A Note on Base Rates and Psychometric Efficiency." *Journal of Consulting Psychology,* 1962, *26,* 422–424.

Dawes, R. M., and Corrigan, B. "Linear Models in Decision Making." *Psychological Bulletin,* 1974, *81,* 95–106.

Derogatis, L. R. *The SCL-90 Manual: Scoring, Administration and Procedures for the SCL-90.* Baltimore: School of Medicine, Johns Hopkins University, 1977.

Dixon, W. J. (Ed.). *BMDP Statistical Software 1981.* Berkeley: University of California Press, 1981.

Duran, B. S., and Odell, P. L. *Cluster Analysis: A Survey.* Berlin: Springer-Verlag, 1974.

Eckart, C., and Young, G. "The Approximation of One Matrix by Another of Lower Rank." *Psychometrika,* 1936, *1,* 211–218.

Edelbrock, C. "Mixture Model Tests of Hierarchical Clustering Algorithms: The Problem of Classifying Everybody." *Multivariate Behavioral Research,* 1979, *14,* 367–384.

Edelbrock, C., and Achenbach, T. M. "A Typology of Child Behavior Profile Patterns." *Journal of Abnormal Child Psychology,* 1980, *8,* 441–470.

Edelbrock, C., and McLaughlin, B. "Hierarchical Cluster Analysis Using Intraclass Correlations: A Mixture Model Study." *Multivariate Behavioral Research,* 1980, *15,* 299–318.

Edwards, A.W.F., and Cavalli-Sforza, L. "A Method for Cluster Analysis." *Biometrics,* 1965, *21,* 362–375.

Einhorn, H. J., and Hogarth, R. M. "Unit Weighting: Schemes for Decision Making." *Organizational Behavior and Human Performance,* 1975, *13,* 171–192.

Ennis, J. G. "Blockmodels and Spatial Representations of Group Structure: Some Comparisons." In Herschel C. Hudson and associates (Eds.), *Classifying Social Data: New Applications of Analytic Methods for Social Science Research.* San Francisco: Jossey-Bass, 1982.

Everitt, B. S. *Cluster Analysis.* London: Halstead Press, 1974.

Everitt, B. S., Gourlay, A. J., and Kendell, R. E. "An Attempt at Validation of Traditional Psychiatric Syndromes by Cluster Analysis." *British Journal of Psychiatry,* 1971, *119,* 399–412.

Feild, H. S., and Schoenfeldt, L. F. "Ward and Hook Revisited: A Two-Part Procedure for Overcoming a Deficiency in the Grouping of Persons." *Educational and Psychological Measurement,* 1975, *35,* 171–173.

Feild, H. S., Lissitz, R. W., and Schoenfeldt, L. F. "The Utility of Homogeneous Subgroups and Individual Information in Prediction." *Multivariate Behavioral Research,* 1975, *10,* 449–460.

Fisher, R. A. "The Use of Multiple Measurements in Taxonomic Problems." *Annals of Eugenics,* 1936, *7,* 179–188.

Fleiss, J. L., and Zubin, Z. "On the Methods and Theory of Clustering." *Multivariate Behavioral Research,* 1969, *4,* 235–250.

Florek, K., and others. "Sur la Liaison et la Division des Points d'un Ensemble Fini." *Colloquium Mathematics,* 1951, *2,* 282–285.

Forgy, E. W. "Cluster Analysis of Multivariate Data: Efficiency Versus Interpretability of Classifications." *Biometrics,* 1965, *21,* 768.

Frank, B. A. "A Comparison of an Actuarial and a Linear Model for Predicting Organizational Behavior." *Applied Psychological Measurement,* 1980, *4,* 171–181.

French, J. W. "Motivational Types Among College Students." *Multivariate Behavioral Research,* 1970, *5,* 135–151.

Friedman, H. P., and Rubin, J. "On Some Invariant Criteria for Grouping Data." *Journal of the American Statistical Association,* 1967, *62,* 1159–1178.

Gaito, J. "Measurement Scales and Statistics: Resurgence of an Old Misconception." *Psychological Bulletin,* 1980, *87,* 564–567.

Gengerelli, J. A. "A Method for Detecting Subgroups in a Population and Specifying Their Membership." *Journal of Psychology,* 1963, *55,* 457–468.

Gilberstadt, H. *Comprehensive MMPI Code Book for Males.* Minneapolis: Veterans Administration Hospital, 1970.

Gilberstadt, H., and Duker, J. *Clinical and Actuarial MMPI Interpretation.* Philadelphia: Saunders, 1965.

Gilmour, J.S.L. "A Taxonomic Problem." *Nature,* 1937, *139,* 1040–1042

Gleser, G. C. "Projective Methodologies." *Annual Review of Psychology,* 1963, *14,* 391–422.

Goldberg, L. R. "Diagnosticians Versus Diagnostic Signs: The Diagnosis of Psychosis Versus Neurosis from the MMPI." *Psychological Monographs,* 1965, *79,* whole no. 602.

Goldberg, L. R. "The Search for Configural Relationships in Personality Assessment: The Diagnosis of Psychosis Versus Neurosis from the MMPI." *Multivariate Behavioral Research,* 1969, *4,* 523–536.

Golden, R. R., and Meehl, P. E. "Detection of Biological Sex: An Empirical Test of Cluster Methods." *Multivariate Behavioral Research,* 1980, *15,* 475–496.

Goldstein, S. G., and Linden, J. D. "A Comparison of Multivariate

Grouping Techniques Commonly Used with Profile Data." *Multivariate Behavioral Research,* 1969, *4,* 103–114.

Goodall, D. W. "Objective Methods for the Classification of Vegetation. I: The Use of Positive Interspecific Correlations." *Australian Journal of Botany,* 1953, *1,* 39–63.

Gordon, L. V., and Sait, E. M. "Q-Typing in the Domain of Manifest Needs." *Educational and Psychological Measurement,* 1969, *29,* 87–98.

Gorsuch, R. L. *Factor Analysis.* Philadelphia: Saunders, 1974.

Gower, J. C. "Some Distance Properties of Latent Roots and Vector Methods Used in Multivariate Analysis." *Biometrika,* 1966, *53,* 325–338.

Gower, J. C. "A Comparison of Some Methods of Cluster Analysis." *Biometrics,* 1967, *23,* 623–637.

Gower, J. C., and Ross, G.J.S. "Minimum Spanning Trees and Single Linkage Cluster Analysis." *Applied Statistics,* 1969, *18,* 54–64.

Green, B. F., Jr. *Digital Computers in Research.* New York: McGraw-Hill, 1963.

Green, B. F., Jr. "On the Factor Score Controversy." *Psychometrika,* 1976. *41,* 263–266.

Green, P. E., and Carmone, F. J. *Multidimensional Scaling and Related Techniques in Marketing Analysis.* Boston: Allyn & Bacon, 1970.

Green, P. E., and Carroll, J. D. *Mathematical Tools for Applied Multivariate Analysis.* New York: Academic Press, 1976.

Green, P. E., and Rao, V. R. *Applied Multidimensional Scaling: A Comparison of Approaches and Algorithms.* New York: Holt, Rinehart and Winston, 1972.

Gross, A. L. "A Monte Carlo Study of the Accuracy of the Hierarchical Grouping Procedure." *Multivariate Behavioral Research,* 1972, *7,* 379–389.

Gross, A. L. "A Comparison of the Predictive Accuracy of a Pooling and a Subgrouping Prediction Strategy." *Multivariate Behavioral Research,* 1973, *8,* 341–355.

Guertin, W. H. "The Search for Recurring Patterns Among Individual Profiles." *Educational and Psychological Measurement,* 1966, *26,* 151–165.

Guertin, W. H., and Bailey, J. P., Jr. *Introduction to Modern Factor Analysis.* Ann Arbor: Edwards Brothers, 1970.

Guilford, J. P. *Fundamental Statistics in Psychology and Education.* New York: McGraw-Hill, 1965.

Haggard, E. A. *Intraclass Correlation and the Analysis of Variance.* New York: Dryden Press, 1958.

Harman, H. H. *Modern Factor Analysis.* (Rev. ed.) Chicago: University of Chicago Press, 1976.

Harris, C. W. "Relations Among Factors of Raw Deviation and Double-Centered Score Matrices." *Journal of Experimental Education,* 1953, *22,* 53–58.

Harris, C. W. "Characteristics of Two Measures of Profile Similarity." *Psychometrika,* 1955, *20,* 289–297.

Hartigan, J. A. "Representation of Similarity Matrices by Trees." *Journal of the American Statistical Association,* 1967, *62,* 1140–1158.

Hartigan, J. A. *Clustering Algorithms.* New York: Wiley, 1975.

Hathaway, S. R., and McKinley, J. C. *Minnesota Multiphasic Personality Inventory: Manual.* (Rev. ed.) New York: Psychological Corporation, 1951.

Hawkins, D. M. "The Detection of Errors in Multivariate Data Using Principal Components." *Journal of the American Statistical Association,* 1974, *69,* 340–344.

Heermann, E. F. "Comments on Overall's 'Multivariate Methods for Profile Analysis.'" *Psychological Bulletin,* 1965, *63,* 128.

Helmstadter, G. C. "An Empirical Comparison of Methods for Estimating Profile Similarity." *Educational and Psychological Measurement,* 1957, *17,* 17–82.

Hempel, G. G. *Symposium: Problems of Concept and Theory Formation in the Social Sciences, Language and Human Rights.* Philadelphia: University of Pennsylvania Press, 1952.

Hendrickson, A. E., and White, P. O. "Promax: A Quick Method for Rotation to Oblique Simple Structure." *British Journal of Statistical Psychology,* 1964, *17,* 65–68.

Hicks, L. E. "Some Properties of Ipsative, Normative, and Forced-Choice Normative Measures." *Psychological Bulletin,* 1970, *74,* 167–184.

Holley, J. W. "On the Generalization of the Burt Reciprocity Principle." *Multivariate Behavioral Research,* 1970, *5,* 241–250.

Holley, J. W., and Guilford, J. P. "A Note on the *G* Index of Agreement." *Educational and Psychological Measurement,* 1964, *24,* 749–753.

Horn, D. "A Study of Personality Syndromes." *Character and Personality,* 1943, *12,* 257–274.

Horst, P. "Pattern Analysis and Configural Scoring." *Journal of Clinical Psychology,* 1954, *10,* 3–11.

Horst, P. *Matrix Algebra for Social Scientists.* New York: Holt, Rinehart and Winston, 1963.

Horst, P. *Factor Analysis of Data Matrices.* New York: Holt, Rinehart and Winston, 1965.

Hubert, L. G., and Levin, J. R. "Evaluating Object Set Partitions: Free Sort Analysis and Some Generalizations." *Journal of Verbal Learning and Verbal Behavior,* 1976, *15,* 459–470.

Hudson, H. C., and associates. *Classifying Social Data.* San Francisco: Jossey-Bass, 1982.

Jancey, R. C. "Multidimensional Group Analysis." *Australian Journal of Botany,* 1966, *14,* 127–130.

Jardine, N., and Sibson, R. "The Construction of Hierarchic and Nonhierarchic Classifications." *Computer Journal,* 1968, *11,* 177–184.

Jardine, N., and Sibson, R. *Mathematical Taxonomy.* London: Wiley, 1971.

Jay, R. L. "Q-Technique Factor Analysis of the Rokeach Dogmatism Scale." *Educational and Psychological Measurement,* 1969, *29,* 453–459.

Johnson, S. C. "Hierarchical Clustering Schemes." *Psychometrika,* 1967, *32,* 241–254.

Kaiser, H. F. "Formulas for Component Scores." *Psychometrika,* 1962a, *27,* 83–87.

Kaiser, H. F. "Scaling a Simplex." *Psychometrika,* 1962b, *27,* 155–162.

Kendall, M. G. *Rank Correlation Methods.* London: Griffin, 1948.

King, B. "Stepwise Clustering Procedures." *Journal of the American Statistical Association,* 1967, *62,* 86–101.

Kohlberg, L. "The Development of Children's Orientation Toward

a Moral Order. I: Sequence in the Development of Moral Thought." *Vita Humana,* 1963, *6,* 11–33.

Kruskal, J. B. "On the Shortest Spanning Subtree of a Graph and the Traveling Salesman Problem." *Proceedings of the American Mathematical Society,* 1956, no. 7, 48–50.

Kruskal, J. B. "Nonmetric Multidimensional Scaling." *Psychometrika,* 1964, *29,* 115–129.

Kruskal, J. B. "The Relationship Between Multidimensional Scaling and Clustering." In J. Van Ryzin (Ed.), *Classification and Clustering.* New York: Academic Press, 1977.

Kruskal, W. H. "Ordinal Measures of Association." *Journal of the American Statistical Association,* 1958, *53,* 814–861.

Kuiper, F. K., and Fisher, L. A. "A Monte Carlo Comparison of Six Clustering Procedures." *Biometrics,* 1975, *31,* 777–783.

Lambert, J. M., and Williams, W. T. "Multivariate Methods in Plant Ecology. IV: Nodal Analysis." *Journal of Ecology,* 1962, *50,* 775–802.

Lance, G. N., and Williams, W. T. "A General Theory of Classificatory Sorting Strategies. I: Hierarchical Systems." *Computer Journal,* 1967, *9,* 373–380.

Lanyon, R. I. *A Handbook of MMPI Group Profiles.* Minneapolis: University of Minnesota Press, 1968.

Lord, F. M. "On the Statistical Treatment of Football Numbers." *American Psychologist,* 1953, *8,* 750–751.

Lorr, M. (Ed.). *Explorations in Typing Psychotics.* New York: Pergamon, 1966.

Lorr, M. "Cluster and Typological Analysis." In P. M. Bentler, D. J. Lettieri, and G. A. Austin (Eds.), *Data Analysis Strategies and Designs for Substance Abuse Research.* Washington, D.C.: National Institute on Drug Abuse, 1976.

Lorr, M. "Note on 'A Comparison of Four Clustering Methods.'" *Applied Psychological Measurement,* 1981, *5,* 73–74.

Lorr, M., and Fields, V. "A Factorial Study of Body Types." *Journal of Clinical Psychology,* 1954, *10,* 182–185.

Lorr, M., and Radhakrishnan, B. K. "A Comparison of Two Methods of Cluster Analysis." *Educational and Psychological Measurement,* 1967, *27,* 47–53.

Lorr, M., and Suziedelis, A. "A Cluster Analytic Approach to MMPI

Profile Types." *Multivariate Behavioral Research,* 1982, *17,* 285–299.

Lorr, M., Jenkins, R. L., and Medland, F. F. "Direct Versus Obverse Factor Analysis: A Comparison of Results." *Educational and Psychological Measurement,* 1955, *15,* 441–449.

Lorr, M., Klett, C. M., and McNair, D. *Syndromes of Psychosis.* London: Pergamon Press, 1963.

Lubin, A., and Osburn, H. G. "A Theory of Pattern Analysis for the Prediction of a Quantitative Criterion." *Psychometrika,* 1957, *22,* 63–73.

Lyerly, S. B. "A Survey of Some Empirical Clustering Procedures." In M. M. Katz, J. O. Cole, and W. E. Barton (Eds.), *The Role and Methodology of Classification in Psychiatry and Psychopathology.* Washington, D.C.: U.S. Public Health Service, 1968.

McClung, J. S. Guilford. "Dimensional Analysis of Inventory Responses in the Establishment of Occupational Personality Types." Doctoral dissertation, University of Southern California, 1963.

McIntyre, R. M., and Blashfield, R. K. "A Nearest-Centroid Technique for Evaluating the Minimum-Variance Clustering Procedure." *Multivariate Behavioral Research,* 1980, *15,* 225–238.

McKeon, J. J. "Hierarchical Cluster Analysis—Computer Programs." Washington, D.C.: Biometric Laboratory, George Washington University, 1967.

McNaughton-Smith, P., and others. "Dissimilarity Analyses." *Nature,* 1964, *202,* 1034–1035.

McNemar, Q. *Psychological Statistics.* (3rd ed.) New York: Wiley, 1962.

MacQueen, J. B. "Some Methods for Classification and Analysis of Multivariate Observations." In L. M. LeCam and J. Neyman (Eds.), *Proceedings of the Fifth Berkeley Symposium on Mathematical Statistics and Probability.* Vol. 1. Berkeley: University of California Press, 1967.

McQuitty, L. L. "Elementary Linkage Analysis for Isolating Orthogonal and Oblique Types and Typal Relevancies." *Educational and Psychological Measurement,* 1957, *17,* 207–229.

McQuitty, L. L. "Typal Analysis." *Educational and Psychological Measurement,* 1961, *21,* 677–696.

McQuitty, L. L. "Rank Order Typal Analysis." *Educational and Psychological Measurement,* 1963, *23,* 55–61.

McQuitty, L. L., and Clark, J. A. "Clusters from Iterative, Intercolumnar Correlational Analysis." *Educational and Psychological Measurement,* 1968, *28,* 211–238.

McRae, D. J. "MIKCA: A FORTRAN IV Iterative *K*-Means Cluster Analysis Program." *Behavioral Science,* 1971, *16,* 423–424.

Marks, P. A., and Seeman, W. *The Actuarial Description of Abnormal Personality: An Atlas for Use with the MMPI.* Baltimore: Williams & Wilkins, 1963.

Marks, P. A., Seeman, W., and Haller, D. L. *The Actuarial Use of the MMPI with Adolescents and Adults.* Baltimore: Williams & Wilkins, 1974.

Marriot, F.H.C. "Practical Problems in a Method of Cluster Analysis." *Biometrics,* 1971, *27,* 501–514.

Meehl, P. E. "Configural Scoring." *Journal of Consulting Psychology,* 1950, *14,* 165–171.

Mezzich, J. E. "Evaluating Clustering Methods for Psychiatric Diagnosis." *Biological Psychiatry,* 1978, *13,* 265–281.

Mezzich, J. E. "Comparing Cluster Analytic Methods." In H. C. Hudson and associates (Eds.), *Classifying Social Data: New Applications of Analytic Methods for Social Science Research.* San Francisco: Jossey-Bass, 1982.

Micucci, J. A. "An Investigation of Linear and Configural Prediction with the MMPI." Doctoral dissertation, University of Minnesota, 1980.

Miller, G. A. "A Psychological Method to Investigate Verbal Concepts." *Journal of Mathematical Psychology,* 1969, *6,* 169–191.

Miller, G. A., and Nicely, P. "An Analysis of Perceptual Confusions Among Some English Consonants." *Journal of the Acoustical Society of America,* 1955, *27,* 338–352.

Milligan, G. W. "Ultrametric Hierarchical Clustering Algorithms." *Psychometrika,* 1979, *44,* 343–346.

Milligan, G. W. "An Examination of the Effect of Six Types of Error Perturbation on Fifteen Clustering Algorithms." *Psychometrika,* 1980, *45,* 325–342.

Milligan, G. W. "A Monte Carlo Study of Thirty Internal Measures for Cluster Analysis." *Psychometrika,* 1981a, *46,* 187–199.

Milligan, G. W. "A Review of Monte Carlo Tests of Cluster Analysis." *Multivariate Behavioral Research,* 1981b, *16,* 379–407.

Milligan, G. W., and Isaac, P. D. "The Validation of Four Ultrametric Clustering Algorithms." *Pattern Recognition,* 1980, *12,* 41–50.

Milligan, G. W., and Sokal, L. M. "A Two-Stage Clustering Algorithm with Robust Recovery Characteristics." *Educational and Psychological Measurement,* 1980, *40,* 755–759.

Mojena, R. "Hierarchical Grouping Methods and Stopping Rules: An Evaluation." *Computer Journal,* 1977, *20,* 359–363.

More, W. W. "Principal Component Analysis to Identify Multivariate Outliers Prior to Clustering." *Classification Society Bulletin,* 1981, *5,* 27.

Morgan, J. N., and Sonquist, J. A. "Problems in the Analysis of Survey Data and a Proposal." *Journal of the American Statistical Association,* 1963, *58,* 415–435.

Morris, J. D. "The Predictive Accuracy of Full Rank Variables Versus Various Types of Factor Scores: Implications for Test Validation." *Educational and Psychological Measurement,* 1980, *40,* 389–396.

Nerviano, V. J. "Common Personality Patterns Among Alcoholic Males: A Multivariate Study." *Journal of Consulting and Clinical Psychology,* 1976, *44,* 104–110.

Nie, N. H., and others. *SPSS: Statistical Package for the Social Sciences.* New York: McGraw-Hill, 1975.

Nunnally, J. C. "The Analysis of Profile Data." *Psychological Bulletin,* 1962, *59,* 311–319.

Nunnally, J. C. *Psychometric Theory.* New York: McGraw-Hill, 1967.

Osgood, C. E., Suci, G. J., and Tannenbaum, P. H. *The Measurement of Meaning.* Urbana: University of Illinois Press, 1957.

Overall, J. E. "Note on Multivariate Methods for Profile Analysis." *Psychological Bulletin,* 1964, *61,* 195–198.

Overall, J. E., and Klett, C. J. *Applied Multivariate Analysis.* New York: McGraw-Hill, 1972.

Owens, W. A. "Toward One Discipline of Scientific Psychology." *American Psychologist,* 1968, *23,* 782–785.

Owens, W. A. "A Quasi-Actuarial Basis for Individual Assessment." *American Psychologist,* 1971, *26,* 992–999.

Owens, W. A. "Background Data." In Marvin D. Dunnette (Ed.), *Handbook of Industrial and Organizational Psychology.* Chicago: Rand McNally, 1976.

Paykel, E. S. "Classification of Depressed Patients—Cluster Analysis Derived Grouping." *British Journal of Psychiatry,* 1971, *118,* 275.

Payne, F. D., and Wiggins, J. S. "The Effect of Rule Relaxation and System Combination on Classification Rates in Two 'Cookbook' Systems." *Journal of Consulting and Clinical Psychology,* 1968, *32,* 734–736.

Peay, E. G. "Nonmetric Grouping: Clusters and Cliques." *Psychometrika,* 1975, *40,* 297–313.

Pinto, P. R. "Subgrouping in Prediction: A Comparison of Moderators and Actuarial Approaches." Doctoral dissertation, University of Georgia, 1970.

Rand, W. M. "Objective Criteria for the Evaluation of Clustering Methods." *Journal of the American Statistical Association,* 1971, *66,* 846–850.

Rao, C. R. *Advanced Statistical Methods in Biometric Research.* New York: Wiley, 1952.

Rice, C. E., and Lorr, M. *An Empirical Comparison of Typological Analysis Methods.* (Contract N00014-67-A-0214.) Washington, D.C.: Office of Naval Research, 1969.

Rohlf, F. J. "An Empirical Comparison of Three Ordination Techniques in Numerical Taxonomy." *Systematic Zoology,* 1972, *21,* 271–280.

Rohlf, F. J. "Methods of Comparing Classifications." *Annual Review of Ecology and Systematics,* 1974, *5,* 101–113.

Rohlf, F. J., Kishpangh, J., and Kirk, D. *NT-SYS Numerical Taxonomy System of Multivariate Statistical Program.* New York: State University of New York at Stony Brook, 1971.

Ross, J. "The Relation Between Test and Person Factors." *Psychological Review,* 1963, *70,* 432–443.

Ruda, E. S. "The Effect of Interpersonal Similarities on Management Performance." Doctoral dissertation, Purdue University, 1970.

Ryder, R. G. "Profile Factor Analysis and Variable Factor Analysis." *Psychological Reports,* 1964, *15,* 119–127.

Sammon, J. W. "A Non-Linear Mapping for Data Structure Analysis." *IEEE Transactions on Computers,* 1969, *18,* 401–409.

Sampson, S. F. *Crisis in a Cloister.* University Microfilms W69-5775. Ann Arbor: University of Michigan, 1969.

Sandler, J. "The Reciprocity Principle as an Aid for Factor Analysis." *British Journal of Psychology Statistical Section,* 1949, *2,* 180–187.

Saunders, D. R., and Schucman, H. "Syndrome Analysis: An Efficient Procedure for Isolating Meaningful Subgroups in a Nonrandom Sample of a Population." Paper presented at 3rd annual meeting of the Psychonomic Society, St. Louis, September 1962.

Schoenfeldt, L. F. "Life Experience Subgroups as Moderators in the Prediction of Educational Criteria." Paper presented at the American Educational Research Association meeting, Minneapolis, 1970.

Sheldon, W. H., Stevens, S. S., and Tucker, W. B. *The Varieties of Human Physique.* New York: Harper & Row, 1941.

Shepard, R. N. "The Analysis of Proximities: Multidimensional Scaling with an Unknown Distance Function." *Psychometrika,* 1962, *27,* 125–139.

Shepard, R. N. "Multidimensional Scaling, Tree-Fitting, and Clustering." *Science,* 1980, *210* (4468), 390–398.

Shepard, R. N., and Arabie, P. "Additive Clustering: Representation of Similarities as Combinations of Discrete Overlapping Properties." *Psychological Review,* 1979, *86,* 87–123.

Shepard, R. N., Kilpatric, D. W., and Cunningham, J. P. "The Internal Representation of Numbers." *Cognitive Psychology,* 1975, *7,* 82–138.

Shepard, R. N., Romney, A. K., and Nerlove, S. B. (Eds.). *Multidimensional Scaling.* Vol. 1. New York: Seminar Press, 1972.

Siegel, S. *Nonparametric Statistics for the Behavioral Sciences.* New York: McGraw-Hill, 1956.

Sines, J. O. "Actuarial Methods in Personality Assessment." In B. A. Maher (Ed.), *Progress in Experimental Personality Research.* Vol. 3. New York: Academic Press, 1966.

Sines, J. O. "Actuarial Prediction." In P. M. Bentler, D. J. Lettieri, and G. A. Austin (Eds.), *Data Analysis Strategies and Designs for*

Substance Abuse Research. Washington, D.C.: Government Printing Office, 1976.

Sjoberg, L., and Holley, J. W. "A Measure of Similarity Between Individuals When Scoring Directions of Variables Are Arbitrary." *Multivariate Behavioral Research,* 1967, *2,* 377–384.

Skinner, H. A. "Differentiating the Contribution of Elevation, Scatter, and Shape in Profile Similarity." *Educational and Psychological Measurement,* 1978, *38,* 297–308.

Skinner, H. A. "Dimensions and Clusters: A Hybrid Approach to Classification." *Applied Psychological Measurement,* 1979, *3,* 327–341.

Skinner, H. A., and Lee, H. "Modal Profile Analysis: A Computer Program for Classification Research." *Educational and Psychological Measurement,* 1980, *40,* 769–772.

Skinner, H. A., and Jackson, D. N. "A Model of Psychopathology Based on an Integration of MMPI Actuarial Systems." *Journal of Consulting and Clinical Psychology,* 1978, *46,* 231–238.

Sneath, P.H.A. "The Application of Computers to Taxonomy." *Journal of General Microbiology,* 1957, *17,* 201–226.

Sneath, P.H.A. "A Comparison of Different Clustering Methods as Applied to Randomly Spaced Points." *Classification Society Bulletin,* 1966, *1,* 2–18.

Sneath, P.H.A., and Sokal, R. R. *Numerical Taxonomy.* San Francisco: W. H. Freeman, 1973.

Sokal, R. R., and Michener, C. D. "A Statistical Method for Evaluating Systematic Relationships." *University of Kansas Science Bulletin,* 1958, *38,* 1409–1438.

Sokal, R. R., and Sneath, P.H.A. *Principles of Numerical Taxonomy.* San Francisco: W. H. Freeman, 1963.

Somers, R. H., and others. "Automatic Interaction Detection." In P. M. Bentler, D. J. Lettieri, and G. A. Austin (Eds.), *Data Analysis Strategies and Designs for Substance Abuse Research.* Washington, D.C.: National Institute on Drug Abuse, 1976.

Sonquist, J. A., and Morgan, J. N. *The Detecting Interaction Effects.* Monograph 35. Ann Arbor: Institute for Survey Research, University of Michigan, 1964.

Sorensen, T. A. "A Method of Establishing Groups of Equal Amplitude in Plant Sociology Based on Similarity of Species Content

and Its Application to Analysis of the Vegetation on Danish Commons." *Biologiske Skrifter,* 1948, *5,* 1–34.

Spath, H. *Cluster Analysis Algorithms.* New York: Wiley, 1980.

Stein, M. I., and Neulinger, J. "A Typology of Self-Descriptions." In M. M. Katz, J. O. Cole, and W. E. Barton (Eds.), *The Role and Methodology of Classification in Psychiatry and Psychopathology.* Chevy Chase: National Institute of Mental Health, 1968.

Stephenson, W. "Introduction to Inverted Factor Analysis with Some Applications to Studies in Orexia." *Journal of Educational Psychology,* 1936, *5,* 353–367.

Stephenson, W. *The Study of Behavior: Q-Technique and Its Methodology.* Chicago: University of Chicago Press, 1953.

Stevens, S. S. "On the Theory of Scales of Measurement." *Science,* 1946, *103,* 677–680.

Suziedelis, A., Lorr, M., and Tonesk, X. "Comparison of Item-Level and Score-Level Typological Analysis: A Simulation Study." *Multivariate Behavioral Research,* 1976, *11,* 135–145.

Tellegen, A. "Direction of Measurement: A Source of Misinterpretation." *Psychological Bulletin,* 1965, *63,* 233–243.

Thomson, G. H. *The Factorial Analysis of Human Ability.* Boston: Houghton Mifflin, 1950.

Toops, H. A. "The Use of Addenda in Experimental Control, Social Census, and Managerial Research." *Psychological Bulletin,* 1948, *45,* 41–74.

Toops, H. A. "A Research Utopia in Industrial Psychology." *Personnel Psychology,* 1959, *12,* 189–225.

Tryon, R. C. *Cluster Analysis.* Ann Arbor: Edwards Brothers, 1939.

Tryon, R. C. *Identification of Social Areas by Cluster Analysis.* Berkeley: University of California Press, 1955.

Tryon, R. C., and Bailey, D. E. *Cluster Analysis.* New York: McGraw-Hill, 1970.

Tucker, L. R. "Factor Analysis of Relevance Judgments: An Approach to Content Validity." *Invitational Conference on Testing Problems.* Princeton, N.J.: Educational Testing Service, 1961.

Tversky, A. "Features of Similarity." *Psychological Review,* 1977, *84,* 327–352.

Van Ryzin, J. (Ed.). *Classification and Clustering.* New York: Academic Press, 1977.

Veldman, D. J. *Fortran Programming for the Behavioral Sciences.* New York: Holt, Rinehart and Winston, 1967.

Wainer, H. "Estimating Coefficients in Linear Models: It Don't Make No Nevermind." *Psychological Bulletin,* 1976, *83,* 213–217.

Ward, J. H., Jr. "Hierarchical Grouping to Optimize an Objective Function." *Journal of the American Statistical Association,* 1963, *58,* 236–244.

Ward, J. H., Jr., and Hook, M. E. "Application of an Hierarchical Grouping Procedure to a Problem of Grouping Profiles." *Educational and Psychological Measurement,* 1963, *23,* 69–81.

Weiss, D. J. "Factor Analysis and Counseling Research." *Journal of Counseling Psychology,* 1970, *17,* 477–485.

Wiggins, J. S. *Personality and Prediction: Principles of Personality Assessment.* Reading, Mass. Addison-Wesley, 1973.

Wilks, S. S. "Weighting Systems for Linear Functions of Correlated Variables When There Is No Dependent Variable." *Psychometrika,* 1938, *3,* 23–40.

Wishart, D. "An Algorithm for Hierarchical Classification. *Biometrics,* 1969a, *25,* 165–170.

Wishart, D. "Mode Analysis." In A. J. Cole (Ed.), *Numerical Taxonomy.* New York: Academic Press, 1969b.

Wishart, D. *CLUSTAN User Manual.* (3rd ed.) Edinburgh: Program Library Unit, Edinburgh University, 1978.

Wolfe, J. H. "Pattern Clustering by Multivariate Mixture Analysis." *Multivariate Behavioral Research,* 1970, *5,* 329–350.

Wolfe, J. H. *Comparative Cluster Analysis of Patterns of Vocational Interest.* Technical Bulletin STB 72-3. San Diego: Naval Personnel and Training Research Laboratory, 1971.

Zahn, C. T. "Graph-Theoretical Methods for Detecting and Describing Gestalt Clusters." *IEEE Transactions on Computers,* 1971, *C-20,* 68–86.

Index

A

B

C

N

O

P

Q

R

T

U

D1271197